T0210728

Springer Aerospace Technology

Series Editors

Sergio De Rosa, DII, University of Naples Federico II, Napoli, Italy

Yao Zheng, School of Aeronautics and Astronautics, Zhejiang University, Hangzhou, Zhejiang, China

Elena Popova, AirNavigation Bridge Russia, Chelyabinsk, Russia

The series explores the technology and the science related to the aircraft and spacecraft including concept, design, assembly, control and maintenance. The topics cover aircraft, missiles, space vehicles, aircraft engines and propulsion units. The volumes of the series present the fundamentals, the applications and the advances in all the fields related to aerospace engineering, including:

- structural analysis,
- aerodynamics,
- aeroelasticity,
- aeroacoustics,
- flight mechanics and dynamics
- orbital maneuvers,
- avionics,
- systems design,
- materials technology,
- launch technology,
- payload and satellite technology,
- space industry, medicine and biology.

The series' scope includes monographs, professional books, advanced textbooks, as well as selected contributions from specialized conferences and workshops.

The volumes of the series are single-blind peer-reviewed.

To submit a proposal or request further information, please contact: Mr. Pierpaolo Riva at pierpaolo.riva@springer.com (Europe and Americas) Mr. Mengchu Huang at mengchu.huang@springer.com (China)

The series is indexed in Scopus and Compendex

Antonio Filippone · George Barakos

Editors

Lecture Notes in Rotorcraft Engineering

 Springer

Editors
Antonio Filippone (iD)
University of Manchester
Manchester, UK

George Barakos
School of Engineering
University of Glasgow
Glasgow, UK

ISSN 1869-1730   ISSN 1869-1749 (electronic)
Springer Aerospace Technology
ISBN 978-3-031-12439-6   ISBN 978-3-031-12437-2 (eBook)
https://doi.org/10.1007/978-3-031-12437-2

This Springer imprint is published by the registered company Springer Nature Switzerland AG
The registered company address is: Gewerbestrasse 11, 6330 Cham, Switzerland

Preface

This book arises from the research activities of the UK Vertical Lift Network (VLN), an organisation that groups together academic researchers, the aerospace industry, small and medium enterprise and the UK Ministry of Defence. In recent years, this network has been channelling all interests in rotor and vertical lift systems under a single long-term strategy initiated by the industry. One of these initiatives has produced a set education and training programmes.

These programmes have been delivered both in person (across Universities in the UK), in the laboratory and by video conference technologies. They ranged from 2-hour sessions to full-immersion two-day events. Since many Universities do not offer specific modules in rotorcraft as part of their degree programmes, the *Network* thought of a collection of notes serving as a first taste of what is a fascinating engineering topic.

In fact, this book is a distillation of the training materials and includes a series of video recordings of computer simulations that can be used alongside the main text. A number of items could not be covered, but it is hoped that additional topics can be added in future.

Each author has contributed according to their expertise and professional interests, and this has resulted in a multidisciplinary compendium that includes several aspects of rotorcraft technology. This book is no substitute for any of the highly regarded textbooks already available in the technical literature. It is to be used alongside the existing literature and may be of interest to anyone wanting to join future training events of the *Vertical Lift Network*.

Manchester, UK
Glasgow, UK
January 2022

Antonio Filippone
George Barakos

Contents

Editors and Contributors

About the Editors

Prof. Antonio Filippone is at the School of Engineering, University of Manchester, where he specialises in aircraft and rotorcraft performance, environmental emissions aircraft noise and engine performance.

Prof. George Barakos is at the School of Engineering, Glasgow University, where he specialises in high-fidelity aerodynamics and aero-acoustics models, computational fluid dynamics, high-performance computing and rotorcraft engineering.

Contributors

Appleton Wesley School of Engineering, The University of Manchester, Manchester, UK

Barakos George School of Engineering, University of Glasgow, Glasgow, Scotland;
Glasgow University, Glasgow, Scotland

Bojdo Nicholas School of Engineering, The University of Manchester, Manchester, UK

Filippone Antonio School of Engineering, The University of Manchester, Manchester, UK

Green Richard School of Engineering, University of Glasgow, Glasgow, Scotland

Morales Rafael School of Engineering, University of Leicester, Leicester, UK

Smith Dale QinetiQ, Rosyth, Scotland

Chapter 1
Rotorcraft Aerodynamics

George Barakos

Abstract This chapter introduces the reader to the aerodynamic aspects of rotary wings. It is written as an introduction for aspiring students, not as a replacement of well-known texts used for rotorcraft classes at University level. It should be used as a first read what is a very complex but fascinating topic in rotorcraft. This chapter illustrates the aerodynamic environment of the rotor, the rotorcraft blade sections and their features, modern aerofoil developments, advanced blade tip designs and advanced rotor design methods. While studying this chapter, one should keep in mind that aerodynamics means different things to different people. When looking at pilot training, the lines between aerodynamics, performance, and aircraft handling are blurred. The technical literature on rotorcraft aerodynamics is dominated by the rotor, but there is a volume of work on fuselage drag as well.

Nomenclature

a_1, b_1	Amplitudes in flap angle equation, Eq. 1.11
A_1, B_1	Amplitudes in blade pitch equation, Eq. 1.5
A	Area
c	Blade chord
C_L	Lift coefficient
C_{Lmax}	Maximum lift coefficient
C_M	Pitching moment
C_T	Thrust coefficient
D	Aerodynamic drag
M	Mach number
M_{tip}	Tip Mach number
r_x	Fraction of the rotor radius

Supplementary Information The online version contains supplementary material available at https://doi.org/10.1007/978-3-031-12437-2_1.

G. Barakos (✉)
School of Engineering, University of Glasgow, Glasgow, Scotland
e-mail: george.barakos@glasgow.ac.uk

R	Rotor radius
R	Gas constant in Eq. 1.14
t	Time
T	Air temperature in Eq. 1.14
U	Air speed
U_p	Out-of-plane (perpendicular) velocity component
U_t	In-plane (parallel) velocity component

Greek Symbols

α	Angle of attack
β	Flap angle
γ	Forward tilt angle of the rotor; ratio between specific heats
λ_i	Inflow ratio
μ	Advance ratio
θ	Pitch angle
θ_o	Collective pitch
ϕ	Inflow angle
ρ	Air density
σ	Rotor solidity
ψ	Azimuth angle
ω	Angular frequency
Ω	Rotor angular speed

1.1   Introduction

This chapter will describe the rotor aerodynamic environment, its physics and flow conditions. It will then move to a historic approach where milestones in our understanding of how rotor aerodynamics work will be briefly presented. Key phenomena dominating rotor flows like dynamic pitch of the blades, variation of Mach number and operation within stall are then to be detailed. Each aspect of the rotor aerodynamics will be addressed using modern results and a combination of evidence from experimentation and simulation. Leaving the sectional aerodynamics, the chapter will progress in the effect of blade planform and how sectional and planform effects are combined to improve rotor performance. There are small practical MATLAB scripts that are used throughout the chapter to allow students to explore the covered effects and develop a better feel of the order of magnitude of the various effects. The chapter closes with a review of modern approached to automate the blade design or employ active structures and element like flaps, tabs, morphing etc.

Rotary wings are used everywhere in aerospace engineering and are nowadays very sophisticated devices designed and optimised with state of the art computer-based methods. At the same time rotary wings are the subject of wind-tunnel inves-

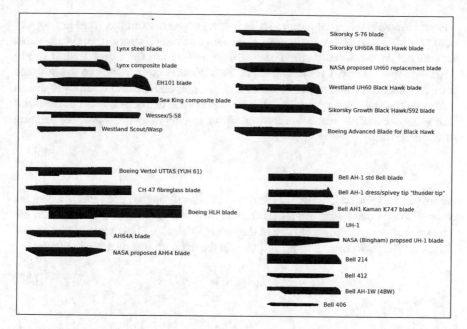

Fig. 1.1 Selection of rotorcraft blade planforms

tigations where very sophisticated flow measurement devices are used to help engineers identify key fluid mechanics phenomena (stall, vortex shedding etc.) that dictate the performance we can obtain when deploying these wings in applications. More interestingly, in recent years, a shift is observed in the research where acoustics is now taking centre stage next to efficiency trying to support the overall objectives of low environmental impact from aircraft.

Typical examples of rotary wing planforms used for helicopters are shown in Fig. 1.1 below where one may observe the lack of a universal trend in design. The main reason for this is the complex aerodynamic environment where rotary wings have to operate within and this will be discussed in the next section.

1.2 The Rotor Aerodynamic Environment

The aerodynamics of a rotor in hover or forward flight is complicated. A useful dimensionless speed which can be used to describe the occurring phenomena is the advance ratio μ defined as:

$$\mu = \frac{U}{\Omega R} = \frac{U}{U_{tip}} \tag{1.1}$$

which is the ratio of flight speed U to tip speed ΩR of a rotary wing of radius R, rotating about a centre of rotation at an angular speed Ω. For most helicopters the advance ratio can reach values up to $\mu = 0.4$. As a typical example, most helicopters operate with tip speeds approximately in the range of 190–220 m/s; $\mu = 0.4$ indicates that the helicopter forward speed of $\sim$80 m/s or 290 km/h for a tip speed of nearly 210 m/s. Most times we are interested in the velocities contributing to the advance ratio in a direction tangential to the rotor disk and perpendicular to it. This projection makes theoretical analyses easier and is used in several textbooks. Most of the times an $x - z$ system of reference is employed with x tangential to the rotor and z pointing through the disk.

$$\mu_x = \mu \cos \gamma, \qquad \mu_z = \mu \sin \gamma \qquad (1.2)$$

where γ is the forward tilt angle of the rotor in radians. Another convention is that the rotor turns anti-clockwise as viewed from above, and the zero of the blade azimuth angle ψ is assumed to be at the back of the disk (6 O'clock). The rotor blade tangential velocity changes from root to tip and around the azimuth during a blade rotation.

$$U_t = U_{tip} [r_x + \mu \sin(\omega t)] \qquad (1.3)$$

$$\psi = \omega t \qquad (1.4)$$

and r_x is the fraction of the rotor radius, varying between 0 (at the centre) to 1 (at the blade tip). Also, t is time and ω is the rate of rotation. The normal velocity U_p of the air through the rotor disk depends on the disk tilt/orientation, the flapping speed of the blades, and the inflow to the rotor. A simple MATLAB script can be used to plot the tangential velocities over the rotor disk (see Fig. 1.2) for a given advance ratio. It is evident that there are different velocities on the advancing (higher velocity) and retreading (lower velocity) sides and as the blade rotates around the azimuth different amounts of lift will be generated on the two sides of the rotor. Consequently, unless the blades pitch and flap as they rotate the rotor alone will not be balanced, trimmed, and will give an unstable and uncontrollable aircraft. Based on the plots shown here, the aircraft will roll to the left.

The most common method to resolve this problem is by trimming the rotor the blades by varying their pitch angle, via what is called the swashplate mechanism, with its associated rotor head assembly that will be discussed later. The result of this mechanism is that we can change the blade pitch in a cyclic way (around the azimuth) according to:

$$\theta = \theta_o - A_1 \sin \psi - B_1 \cos \psi \qquad (1.5)$$

which means that the pitch of the blade is made out of a common (collective) angle with a lateral (sine) cyclic term and a longitudinal (cosine) term.

At the moment, consider a rotor which is trimmed approximately, with forwards disk tilt caused by longitudinal cyclic pitch. Here we have

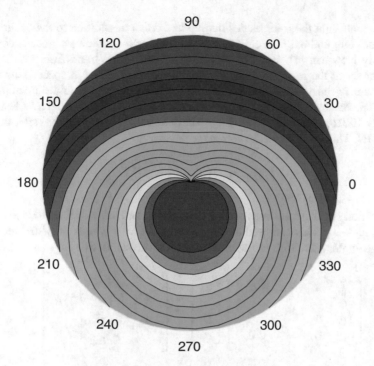

Fig. 1.2 Tangential velocity contours on a rotor disk in forward flight

$$\theta = \theta_0 + B_1 \sin \psi \tag{1.6}$$

The rolling moment coefficient of a rotor in forward flight can be derived from momentum and blade element theories [1]) as

$$C_{MR} = 0.5\sigma a \left[\frac{2}{3} \theta_0 \mu - \frac{B_1}{4} \left(1 + \frac{3}{2} \mu^2 \right) \right] \tag{1.7}$$

The lift curve slope a is empirically estimated to 5.7/radian for rotor blades. For zero rolling moment if forward flight this gives a first estimate of the cyclic pitch angles needed to trim the rotor. Maximum blade pitch in occurs at 270 °C azimuth. For high forward speed (assume here $\mu = 0.4$) the maximum pitch can be set close to the blade section stall angle to exploit the maximum performance of the blade section. If a typical static stall angle for a around 15 °C is used we have:

$$15 = \theta_0 + B_1 \tag{1.8}$$

and setting the moment to zero while eliminating the collective from the above equation gives approximately $\theta_0 = 7.5$ degrees and for the longitudinal cyclic pitch we now have $B_1 = 6.5$ degrees. Using these approximations for a rotor in forward flight balances the tendency of the rotor to roll and results in the lift contours shown below.

In actual flight the rotor blades flap up and down in addition to the cyclic pitch. A typical longitudinal trim calculation process for a medium weight helicopter is given by Newman [1]. For a given all up weight (AUW) and fuselage drag at a given airspeed the resultant main rotor thrust vector is fixed in space. This allows calculation of the disk tilt and fuselage tilt angles for moment equilibrium in forward flight. The fuselage drag is available from wind tunnel tests at a reference test speed (usually 100 ft/s or 30.48 m/s). For a medium helicopter this is of the order of D100 = 1000 N. This is scaled to the correct speed using:

$$D_F = D_{100} \left(\frac{U2}{U_{100}^2} \right) = \frac{1000\, U^2}{30.48^2} \tag{1.9}$$

An iteration is performed between the rotor thrust and inflow values to give C_T and $\mu_{zD} = \mu_z + \lambda_i$. This can be done using the MATLAB scripts provided here. The flapping angles (disk tilt), inflow and rotor thrust are input to

$$\begin{bmatrix} \frac{\gamma}{8} & 0 & -\frac{\gamma}{6}\mu_x & -\lambda_{\mu_{zD}}^2 \\ 0 & -\frac{\gamma}{8} & 0 & -\frac{\gamma}{6}\mu_x \\ \frac{\gamma}{3}\mu_x & 0 & -\frac{\gamma}{8} & 0 \\ \frac{1}{3}\mu_x & 0 & -\frac{\mu_x}{2} & 0 \end{bmatrix} \begin{bmatrix} \theta_0 \\ A_1 \\ B_1 \\ a_0 \end{bmatrix} = \begin{bmatrix} \frac{\gamma}{6}\mu_{zD} \\ -\frac{\gamma}{8}b_1 + \left(1 - \lambda_\beta^2\right)a_1 \\ -\frac{\gamma}{8}a_1 - \left(1 - \lambda_\beta^2\right)b_1 \\ \frac{C_T}{a\sigma}\frac{\mu_{zD}}{2} \end{bmatrix} \tag{1.10}$$

and this may be used to calculate the collective and cyclic angles, together with the rotor coning angle required for longitudinal trim [2]. In this instance, $\mu_{zD} = \mu_z + \lambda_i$ represents the sum of the climb ratio and downwash. A lateral trim analysis is also necessary due to the amount of cyclic pitch necessary to trim out the effects of tail rotor thrust. In the above, a typical set of data would be $\gamma = 7$, for the blades, that is the Lock number, which is a ratio of aerodynamic to flapping inertia; and $\lambda_{\beta^2} = 1.2$ for a semi-rigid hub, is the non-dimensional flapping frequency, which depends on the rotor hub stiffness and design. A sample longitudinal trim calculation in the speed range of $\mu = 0$ to $\mu = 0.35$ is shown in Fig. 1.3.

With forward speed the fuselage rotates forwards and longitudinal cyclic pitch increases to overcome increasing drag. Of note is the fact that the collective pitch decreases from its hover value to a minimum at around 30 m/s; this is evidence that a rotor in forward flight benefits from "translational lift", which is greater than the lift generated in the hover. However, eventually the increased down-flow through the rotor caused by the forwards disk tilt causes the collective to rise at higher speeds. The longitudinal trim equations are strongly coupled to the lateral trim equations. In lateral trim the procedure is essentially the same, but the tail rotor thrust is also important. These changes in flapping, collective and cyclic angles mean that the rotor blades themselves experience rapid changes in angle of attack as they rotate about the azimuth.

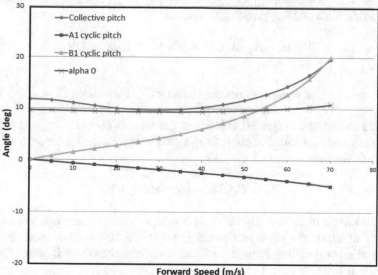

Fig. 1.3 Example of longitudinal rotor trim

1.2.1 *Figure-of-Eight Diagram*

To determine the aerodynamic environment experienced by a rotor blade in forward flight, a 'sausage plot' is used (sometimes called a 'figure of eight' diagram). This is a plot of blade incidence versus Mach number and shows the range of conditions experienced by the rotor at each station. The blade experiences a very severe aerodynamic environment in forward flight. The advancing blade Mach number towards the tip is very high and in the transonic regime while the retreating blade incidence is close to, or above, the static aerofoil section stall angle. This environment leads to conflicting requirements for the selection of an aerofoil, or for the design of a new rotor section. The high advancing side Mach numbers require thin uncambered sections to prevent drag divergence due to compressibility effects (shock formation). The high lift on the retreating side needs good performance at high angles of attack. The best shape for this task is a thick cambered section but this is not easily achievable. Other design aspects of the helicopter also have conflicting requirements (e.g. twist is desirable for good hover performance, but undesirable in forward flight, or large anhedral at the tip of the rotor blade for hover, and low for forward flight). Moreover, blade aeroelasticity should also be taken into account because blades are relatively flexible in torsion and the aerofoil section must therefore have very low pitching moment. The challenge is to design a blade which has good performance in both of these high Mach and high incidence regimes. These design requirements, and aerofoils which meet them, are discussed next.

To generate a figure-of-eight diagram, one has to start from the basic equations for the blade pitch and flap angles given below.

$$\theta = \theta_0 - A_1 \cos \psi - B_1 \sin \psi, \qquad \text{Blade pitch} \qquad (1.11)$$

$$\beta = \beta_0 - a_1 \cos \psi - b_a \sin \psi, \qquad \text{Flap angle} \qquad (1.12)$$

As stated earlier, the tangential velocity at the rotor is $U_t = U_{tip}[r_x + \mu \sin(\omega t)]$ and, we must add the effect of blade flapping to the expression for the perpendicular velocity component as:

$$U_p = U_{tip} [\lambda_i + \mu \beta \cos(\psi)] + r\dot{\beta} \qquad (1.13)$$

The inflow angle of air to the blade is $\tan \phi = U_p/U_t$, and we can use ϕ to estimate the angle of attack at a blade section using $\alpha = \theta - \phi$. At the same time, the Mach number at a blade station located a a fraction r_x from the centre is given by $M = M_{tip}(r_x + \mu \sin(\omega t))$.

For a given station r_x and given the properties of the rotor needed for trimming, and after estimating the downwash λ_i using iterative computations we can obtain the sectional angle of incidence α. A plot of the (M, α) will give the "sausage plot" or "figure of eight plot". A typical high speed plot for the Lynx helicopter is shown in Fig. 1.4.

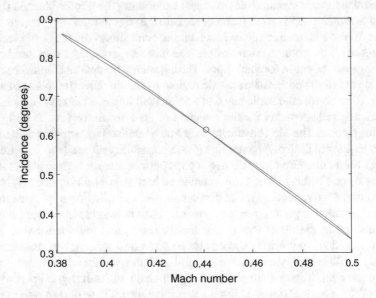

Fig. 1.4 Mach incidence variation of a rotor section at $y/R = 0.75$ and $\mu = 0.1$

1.3 Rotorcraft Aerofoil Sections—History and Current Trends

In the UK, Peter Wilby is perhaps the father of the rotorcraft aerofoil studies. His work included aerofoil design, shape representation aerodynamic performance studies and extensive testing of rotorcraft aerofoils using different wind tunnels. His 1997 Cierva lecture [3] remains relevant today and is a recommended starting point on the topic along with his 1996 paper presented at the European Rotorcraft Forum [4] or his comprehensive work on Vertica [5].

The environment experienced by rotor blades in high speed, forward flight can be summarized by very high incidences, which approach aerofoil stall, on the retreating side and very high Mach numbers on the advancing side. It is these two aerodynamic phenomena, aerofoil stall and shock wave formation, that impose a limit to the forward speed of any helicopter. Rotor blades are high aspect ratio structures. They benefit from a large amount of centrifugal stiffening in blade bending (and flapping motion) but the blade flexibility in torsion is relatively low. Pitching moments associated with stall, transonic effects, or normal rotor operation must be avoided if the blades are not to twist in flight. Initially, symmetric airfoils were used in rotor design and this was by large influenced by the autogyro development where Juan de la Cierva was the first person to use a cambered airfoil section.

A crash in 1939 was blamed on the pitching moments generated by the employed section. Combined with low torsional blade stiffness of early rotors meant that the use of symmetric sections was almost universal until the 1960s. It was only in the 1970s s when progress with computer-based methods allowed for more detailed aerodynamic studies to be performed. Panel methods were first used to allow for iterative studies and implementation of conformal mapping to computer codes was also pivotal. Re-introducing cambered sections was only a matter of time, and this was accelerated by tools allowing the study of transonic flow effects. Ambition with design in the USA and the advent of some high performance helicopters made the aerofoil selection a hot topic of research in the late 1970s. As an example the YAH-64 aircraft that was initially planned to have NACA sections on its blades employed a very advanced, for the time, Hughes HH-02 blade section. (Fig. 1.5)

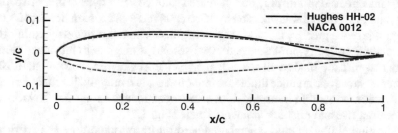

Fig. 1.5 An improved aerofoil section over the NACA series for the blade of the YAH-64 aircraft was the Hughes HH-02 compared with the NACA 23012

Fig. 1.6 Long pitch-links used on rotor heads. The pitch links will suffer higher loads for high-moment aerofoils used on the rotor. Credits: http://www.grubby-fingers-aircraft-illustration.com/

The loss of lift occurring during retreating blade stall is not a major problem. At high forward speed, where retreating blade stall occurs, most of the rotor lift is generated over the fore and aft sectors of the rotor disk at around 60–90% rotor radius so a loss in lift on the retreating blades does not degrade rotor performance significantly. However, there is a large pitching moment change associated with aerofoil stall. This pitching moment change places a large oscillatory load on the blades and their pitch control mechanism. The pitch control linkages may be damaged, or their fatigue life decreased, by retreating blade stall. As a result of these pitching moment loads, retreating blade stall has to be avoided up to as high an advance ratio as possible (Fig. 1.6).

There are two strategies for the aerofoil designer to delay blade stall. One is to increase the blade chord (and hence rotor blade area) and this will decrease the amount of collective pitch, and resulting maximum blade incidence, required to trim the rotor. This strategy can only be used within reason because increased blade chord results in increased profile drag and increased rotor power requirements. Also, blade mass, and the required transmission system mass, increases too. In helicopters rotor design, one of the primary factors for selecting blade chord is to avoid retreating blade stall at VNE (velocity-not-to-exceed). The second strategy is to introduce aerofoil camber so that the blade will reach higher incidences before moment stall. Blade camber is, however, an undesirable feature on the advancing blade where transonic effects are important. In transonic flow, camber can induce strong shock waves with a large drag rise and pitching moment change (Fig. 1.7).

Following Wilby's work, a cambered aerofoil in transonic flow as experienced by the advancing blade will show large pitching moment changes. In transonic flow, the airflow over the upper surface of any aerofoil accelerates to supersonic velocity

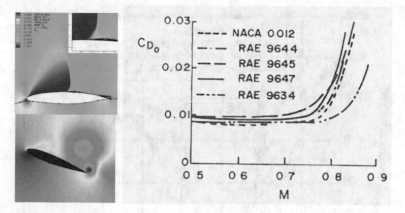

Fig. 1.7 The demanding advancing and retreating side aerodynamics used on the rotors requires sections that can handle Mach flows (outboards where shock waves form) and high-α. Graph on the right from Wilby [5]; graph on the left from Barakos et al. [6]

and forming shock wave recovers to free stream pressure at the trailing edge. The strength and position of this shock wave has a major influence on the amount of transonic drag rise and pitching moment change experienced by the aerofoil. At high forward speed the advancing blade tip is at almost zero incidence (and zero lift). A symmetrical section at zero lift at a Mach number $M = 0.8$ has zero pitching moment, but a cambered section has a strong nose down pitching moment (Fig. 1.8).

The effects of this large, camber-induced, pitching moment on the advancing blade tips can be severe and so the rotor blade and section is designed to minimize the root pitching moment on the advancing blade. The next figure illustrates the problem and shows the torsional load (represented by $M^2 C_m$) for a cambered and a symmetrical tip aerofoil around the azimuth (Fig. 1.9).

It is evident that the cambered section will produce a large once-per-rev oscillatory load each time the advancing blade passes $90\,°C$ azimuth. The loads are very high, despite the fact that aerofoil stall is not encountered in this example. It is for this reason that the symmetric NACA 0012 section was used for many years as the basic rotor blade section. This aerofoil, albeit simple, has reasonably high stall incidence, low pitching moment and tolerable transonic effects.

1.3.1 Modern Aerofoil Developments

The NACA 0012, and NACA 23012, form the basis of several modern rotor blade sections developed by the National Physics Laboratory (NPL) and the Royal Aircraft Establishment (RAE) in the UK. If a basic NACA 0012 section is used with an extended and drooped leading edge as a means to introduce camber then the resulting aerofoil (e.g. the NPL 9615) can give a 10% increase in C_{Lmax} with a small increase

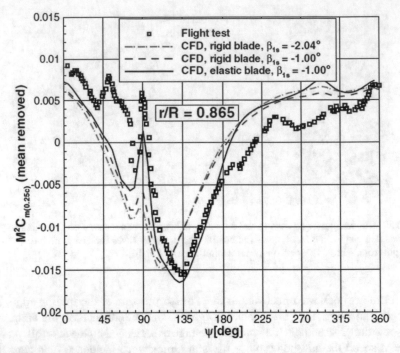

Fig. 1.8 Variation of the blade sectional moment around the azimuth. The figure compares detailed experiments for the UH60A aircraft with simulations using the HMB3 flow solver [7]

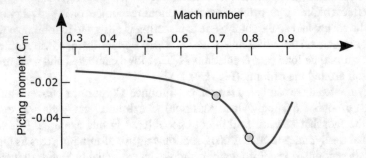

Fig. 1.9 Typical variation of sectional pitching moment coefficient C_m with the Mach number

in Mach number and an acceptable pitching moment at zero lift. Essentially, the nose droop allows an increase in camber without increasing the curvature of the upper surface of the aerofoil and hence avoids the increase in upper surface shock wave strength normally associated with aerofoil camber. The allowable amount of nose droop using this design strategy is limited by the nose down moment that occurs from the underside suction peak (and possible shock wave) resulting from the droop. The introduction of aerofoil camber by this method is illustrated in Fig. 1.10.

Fig. 1.10 Photograph of the RAE 9648 section. ©Getty Images, Science and Society Picture Library

As reported in Wilby's Cierva lecture, research into what pitching moments are sustainable on a rotor was undertaken by the UK RAE on a Wessex helicopter fitted with modified blades (roughened leading edges over certain sections of the blade span). These innovative flight tests at RAE Bedford demonstrated that C_{Lmax} over the inner sections (0 to 60%) of the blades of a helicopter is not a major factor in determining rotor performance and setting the helicopter flight envelope. This is a useful design guideline, because it means that an aerofoil section with reflex camber (an upwards sloping trailing edge for example) and a nose up pitching moment but with poor C_{Lmax} can be used on the inner section of the blades to balance the nose down moment generated by high-lift cambered aerofoil sections further outboard. The modern blade design strategy is therefore to incorporate a modest amount of blade camber on outboard blade sections to delay retreating blade stall, while maintaining low pitching moments and control loads with the use of reflex camber on inboard sections. The NPL 9615 is an example of an early 'modern' rotor aerofoil achieving high lift but with good transonic behaviour. Further research has led to even better section designed by a combination of experience, tunnel testing (at ARA Bedford, and Glasgow University) and numerical methods at RAE and Westland Helicopters (indicial methods by Beddoes) in the RAE 9645 section which is used as the main lifting aerofoil on the Lynx helicopter.

1.3.2 BERP Rotor Aerofoils

The result of many years of tunnel testing (using oscillatory pitching tests to reflect the changing incidences met by rotor blades) and development from the original RAE/NPL 9615 section is a set of three aerofoils each used for a different purpose on the Lynx BERP rotor. RAE 9645 is the main lifting aerofoil used on the Lynx helicopter and is employed between 65 and 85% blade span. This aerofoil is used to achieve the best possible retreating blade stall behaviour. To determine the section's retreating blade stall behaviour, the section is tested in conditions similar to those found on a retreating Lynx blade. A close approximation to these conditions is obtained in a wind tunnel at $M = 0.3$, and where the aerofoil model undergoes sinusoidal pitching 30 Hz and 8 °C half-amplitude. This frequency is selected to obtain a similar reduced frequency $\omega c / U$ between test conditions and real life. An aerofoil undergoing such oscillations experiences dynamic stall—where the change in pitching moment can be as much as 500% of the static stall moment break. The magnitude of the pitching moment break is plotted against the maximum aerofoil incidence in a cycle and the resulting intercept with the incidence axis is a measure of the maximum achievable blade incidence on the rotor.

This aerofoil has a moderate nose down moment at zero lift resulting from its 'nose droop' camber and so its pitching moment on the rotor blade is balanced by a RAE 9648 section inboard, from the hub up to 65% rotor radius. The RAE 9648 is also 12% thick but has a reflexed trailing edge providing a nose up moment. There is a penalty in using this aerofoil at high Mach numbers because of the lower critical Mach number introduced by reflex camber- however, this is not as important for the inboard sections because their Mach numbers are not too high, even at high advance ratio. Finally a tip aerofoil section, the RAE 9634 is used between 85 and 100% span. This aerofoil is designed to delay transonic flow effects and, consequently, it is only 8.3% thick. Sweeping the tip is another strategy used on modern helicopters to delay transonic effects and this along with the overall planform will be discussed in the next section.

1.4 Planform Effects on Rotor Performance

Following from the aerofoil selection, the rotor planform is now the focus of this paragraph. The rotor can always be seen as a wing in constant rotation and lessons learnt from fixed wings can at least partially be applied to rotary wings. Having mentioned the problems arising due to the drag associated with the formation of shock waves, and considering that the outer part of the rotor will suffer more from compressibility effects due to its higher Mach number, the concept of swept blade tips was first explored by rotor designers. Much the same way the swept wing is used in fixed wing aircraft, applying blade tip sweep can help alleviate some of the problems due to shock formation. However, the alleviation of the shock comes with some drawbacks. First of all the centre of pressure shifts rearwards, and the same happens for the inertial axis the mass of the blade tip. This results in a strong nose-down moment that must somehow be dealt with. One mitigation is to apply moderate amounts of sweep and this may work because the advancing side Mach numbers are not very high. This can be observed in Fig. 1.11, where the AW139 and the Boeing AH64A blade tips are shown. Sweep also results in poor retreating blade performance due to early stall.

A typical aerodynamic envelope for a rotor can be seen in the following figure in the form of a loading versus speed diagram. Clearly, there are two limits to the rotor due to the stall on the retreating side and the compressibility effects of the advancing side. As a result the flight envelope is not significantly improved. It is rather stretched to higher speed but not essentially expanded. This is the reason that many radical ideas for planform shapes have been explored using both CFD and experiments (Fig. 1.12).

(a) AW139   (b) AH64-A

Fig. 1.11 Blade tips of the AW139 and AH64-A helicopters. Credits: **a** https://b-domke.de/AviationImages; **b** https://www.airforce-technology.com

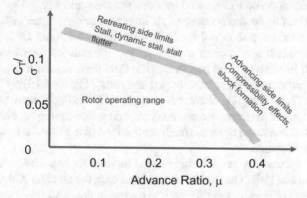

Fig. 1.12 Loading versus advance ratio plot showing the advancing and treating side limitations for conventional rotors

Simply adding sweep will push the speed higher but this will come with reduced lifting capability. A good example of planform design is the British Experimental Rotor Project (BERP). It was a very successful UK (government and industry) programme to develop a much more sophisticated swept tip which would (uniquely) perform well at both high Mach and high incidence. It has a non uniform sweep, incorporates extra frontal area (which counters the aft movement of the centre of pressure), and has a very highly raked tip.

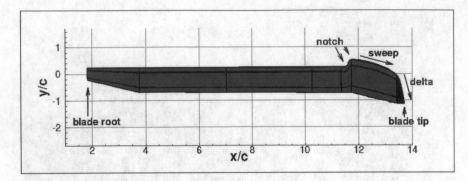

Fig. 1.13 Definition of the parameters of the BERP-like blade

1.4.1 BERP Blade

The BERP blade has a "paddle-shaped" tip due to a forward displacement of the planform (creating a notch) followed by sweep as shown in Fig. 1.13.

It is a benchmark design because of its improved performance in forward flight resulting in a world speed record for helicopter flight [8]. Some CFD analysis was carried out for this blade in order to find the specific reasons for its improved performance, and to analyse what happens in the flow field around it [9]. However, at the time, computational power was limited and today, higher fidelity CFD methods are available. The aim of this chapter is to apply CFD and optimisation methods to investigate if further performance can be obtained by fine-tuning specific features of the BERP tip. At the same time, the methods outlined in the literature survey chapter of this book are applied.

The BERP tip was designed for high speed forward flight without compromising hover performance [10]. The problem associated with the fast forward flight regime, is that the effects of compressibility such as transonic flow and shock waves become significant especially on the advancing blade. Typically, thin aerofoils are used but these tend to stall more easily at the high angles of attack which occur on the retreating side. The first step in the design of the BERP was the aerofoil selection. The aerofoils were selected such that thinner sections could be used to enable higher forward flight speeds. Camber was introduced to improve the stall capability of the blade on the retreating side and the increased pitching moments were alleviated by having a reflexed aerofoil inboards. The resulting blade is reported to behave well in terms of control requirements and twist loads [11].

The planform was then optimised to reduce high Mach number effects by first sweeping the tip of the blade back. This moved the aerodynamic centre of the swept part backwards, causing control problems in the pitch axis. To counteract this, the swept part was translated forward which introduced a notch on the leading edge of the blade. The notch corners were smoothed to avoid flow separation. A "delta" tip was also incorporated so that a stable vortex formed at higher angles of attack on the retreating side to delay stall [9].

One of the characteristics of the blade is that blade stall occurs first inboards of the notch and does not spread outwards. This is because at high angles of attack, such as at the retreating side, the vortex formed travels around the leading edge and the flow over the swept part remains attached. The BERP blade shows similar performance to a standard rotor blade at low speed flight, but superior performance in forward flight due to the absence of drag rise and flow separation [9]. In hover, the Figure of Merit (FM) was improved due to the minimisation of blade area and overall, there were no penalties in hover performance. At high speeds, blade vibration was also reduced as well as control loads for manoeuvres [9].

BERP(3) is currently in use on the Lynx and Merlin helicopters. J. Perry was the main driver behind this project that resulted in a "paddle" tip at the end of the blades. At first the twist of the blades was not uniformly applied but was stronger near the tips to provide the necessary off loading without resulting to too high pitch angles inboard. Then the tip was swept in a non-uniform way so that in hover the Mach number is kept constant along the tip. The shape near the tip is almost like a delta wing, highly raked, and inboard a notch was created increasing the chord and forming the paddle shape. The notch also helps offset the CP aft movement resulting in a more balanced shape with lower pitching moments. At high pitch the notch can also result in an additional vortex helping the tip to stay free from stall at high loading and α.

This design helps so that tips do not stall until a high angle and shock waves are delayed to higher speed. The selection of the tip speed was dictated by a requirement to keep the Mach number $M < 0.88$ at the advancing side to stay below the drag divergence of the employed tip sections.

The advance ratio was also kept below 0.4 so that dynamic stall effects can be tolerable. In hover, the Mach had to be less than 0.7 to avoid shock waves and divergence of drag. Lower limits of speed were set by auto-rotation considerations; this gave a speed of 210 m/s at 160 kt.

The tip speed was decided using several criteria. To begin with,

$$U_{tip} < M_{crit}\sqrt{\gamma RT} - U_{flight} \tag{1.14}$$

In Eq. 1.14, $\gamma = 1.4$, R is the gas constant and T is the absolute temperature. This means that the tip Mach number is below the critical Mach number for the aerofoil used at the tip of the blade. At the same time the dynamic stall of the retreating side should be controlled and this means restricting the advance ratio to about 0.4, or $0.4U_{tip} > U_{flight}$. The hover tip Mach number is also kept low so that the blades operate near max incidence with no shock wave development. This is around 0.68, or $U_{tip} < 0.68\sqrt{\gamma RT}$. The restrictions lead to a map like the one presented in Fig. 1.14.

The blade chord is a difficult selection given that blade area is needed so that stall is avoided especially near the maximum advance ratio. For this case, rotor stability is very important and at the same time the designer has to consider an advance ratio

Fig. 1.14 Tip-speed
considerations during design

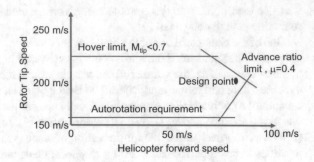

that is above limits set by High pitch link loads, stall flutter, and stability. Thus, lowering the blade incidence requires increasing the blade chord. The blade are is usually selected so that maximum C_T/σ is not exceeded.

$$A_{blade} = \left(\frac{C_T}{\sigma}\right)^{-1} \frac{W}{\rho\,(\Omega R)^2} = 10\,\mathrm{m}^2 \tag{1.15}$$

Selecting different disk loading and specific power loading is also important and the take-off where power is high. Hence, selecting a disk loading and according to the engine characteristics we find out the intersection of power loading vs disk loading with the take-off requirement.

The selection of the number of blades is also interesting. More blades lead to less vibration so dynamics and vibration of the rotor is crucial At the same time good manoeuvrability leads to stiff hubs. Keeping the complexity of the hub low is also important so more than 5 blades increases the complexity. It appears that 4 or 5 blades is the optimum for medium helicopters with light helicopters at times even opting for 3 blades. More blades are usually seen in very heavy helicopters.

1.5 Modern Rotor Design, Methods and Approaches

The design of rotor blades is complex, in that it involves many disciplines of engineering such as aerodynamics, structural dynamics, aeroelasticity and flight control systems. These disciplines do not just play an individual part in the design of the rotor blade, but are coupled; some more strongly than others [12]. Even within a single discipline, such as the aerodynamics of the rotor, there are often conflicting design requirements; Forward flight tends to have opposite requirements to hover, the blade on the advancing side has opposite requirements to that on the retreating side of a forward moving helicopter and so on [13]. Therefore, defining an optimum blade tends to be a compromise between these various conditions. Hence, the optimum is determined by the objectives set for each particular rotor design. While the initial rotor design may be relatively easy to come up with, finding the optimum design parameters is not an easy task.

For the named reasons, computer codes to aid helicopter designers have been, and are still being, developed and used in the industry [14]. In the past, these methods were limited to using simple theories that modelled the aerodynamics of a rotor. This limitation was due to the high cost in obtaining sufficient data to make valid comparisons for a set of design parameters, either by experiments or by simulation. Both methods involve high costs. The computational costs, however, can be reduced by reducing the complexity of the models used to simulate the aerodynamics around a rotor. However, this compromises the accuracy of the data. Nevertheless, over time, computing power has increased allowing more advanced simulation models to be used. This has attracted a lot of interest in the research of design and optimisation methods.

A variety of methods have been developed, and the majority of the applications have been for cases that require a small computational domain (such as aerofoils) or a simple aerodynamic analysis (such as cases where operation is optimised for a single static condition). The challenge is to apply these methods to a complex design such as a helicopter rotor blade, where the aerodynamics are complex and change, and the design space has a large number of dimensions, and to do this accurately and efficiently. The initial concept design is a well-established process and designers and engineers of rotor blades have substantial experience to lean on, as well as the assistance of efficient codes to aid them in obtaining a preliminary rotor design. The optimisation methods become more applicable when optimisation of an existing design is required to obtain even better performance from a rotor.

Optimisation techniques usually require a starting design point, so the design stage is just as important as the optimisation. While it is possible for optimisation to lead to new designs, the time and effort involved would attract a high cost. Take the BERP tip blade as an example [10]. The BERP tip blade is not something that can easily be created by using optimisation methods. However, if the designer's ideas and experience in the field of aerodynamics was used to create an initial design, then the dimensions and extent of the BERP blade characteristics can be perfected to improve performance. When many characteristics of a rotor are adjusted in such a way, a considerable amount of improvement can be made [13]. This is what makes optimisation so important and has led to the increase in the amount of research conducted and the number of academic papers written on this topic.

Alongside the added performance gained by using optimisation procedures, the use of numerical solutions for the optimisation problem removes some of the work-load of obtaining the optimum design from the designer while still giving the designer flexibility. In the case of rotors specifically, there are many criteria and objectives that must be fulfilled simultaneously, what is known as multi-objective optimisation. Depending on the required performance, it is possible to program the optimiser to create a rotor tailored to its expectations in many diverse conditions.

The challenge of optimisation for helicopters is summarised in the following quote from Ref. [15]:

"Researchers in helicopter applications of optimisation face a complex multi disciplinary problem, with several possible choices of design variables, objective functions, behavior and side constraints, analysis models, sensitivity formulations, approximation concepts, optimisation algorithms, not to mention the many types of results that can be generated and presented."

References

1. Newman S (1994) Foundations of helicopter flight. Arnold, Great Britain
2. Seddon J (1990) Basic helicopter aerodynamics. BSP Professional Books, Great Britain
3. Wilby PG (1998) Shockwaves in the rotor world–a personal perspective of 30 years of rotor aerodynamic developments in the UK. Aeronaut J 3:113–128
4. Wilby PG (1996) The development of rotor Aerofoil testing in the UK—the creation of a capability to exploit a design opportunity. Proc Eur Rotorcraft Forum, (Paper no 18):18–1–18–11
5. Wilby PG (1980) The aerodynamic characteristics of some new RAE blade sections, and their potential influence on rotor performance. Vertica 4:121–133
6. Barakos GN, Drikakis D (2003) Computational study of unsteady turbulent flows around oscillating and ramping Aerofoils. Int J Fumer Methods Fluids 42:163–189. https://doi.org/10.1002/fld.578
7. Steijl R, Barakos GN, Badcock KJ (2008) Computational study of the advancing-side lift-phase problem. J Aircraft 45(1):246–257
8. Perry FJ, Wilby PG, Jones AF (1998) The BERP rotor–how does it work, and what has it Beein doing lately. Vertiflite 44(2):44–48
9. Brocklehurst A, Duque EPN (1990) Experimental and numerical study of the british experimental rotor programe blade. In: AIAA 8th applied aerodynamics conference, vol AIAA-90-3008, Portland, Oregon, USA
10. Harrison R, Stacey S, and Hansford B (2008) BERP IV the design, development and testing of an advanced rotor blade. In: American helicopter society 64th annual forum, Montreal, Canada
11. Robinson K, Brocklehurst A (2008) Berp iv—aerodynamics, performance and flight envelope. In: 34th European rotorcraft forum, Liverpool, UK
12. Caradonna FX (1992) The application of CFD to rotary wing problems. In: NASA Technical Memorandum
13. Gordon Leishman J (2005) Principles of helicopter aerodynamics. Cambridge Aerospace Series, 2nd edn
14. Johnson W (2010) NDARC—NASA design and analysis of rotorcraft validation and demonstration. In: American Helicopter society aeromechanics specialist conference, San Francisco, California
15. Celi R (1999) Recent applications of design optimization to rotorcraft—a survey. In: 55th annual forum of the American helicopter society

George Barakos is at the School of Engineering, Glasgow University, where he specialises in high-fidelity aerodynamics and aero-acoustics models, computational fluid dynamics, high-performance computing and rotorcraft engineering.

Chapter 2
Experimental Methods for Aerodynamics

Richard Green

Abstract Experimental techniques for aerodynamics have been essential tools for the development of rotorcraft. This includes wind tunnel, water tank testing, and flight testing. Rotor flows are rich in their physical complexity, and consequently an awareness of range of experimental methods is required. The chapter contains an introduction to wind tunnels, followed by a description of important experimental techniques, such as flow visualisation, pressure measurements, force and moment measurements, thermal anemometry and instrument calibration. Their performance and relative merits are discussed. Results are presented include forces and moments on a rotor, dynamic stall, particle-image velocimetry of a vortex ring state, and more.

Nomenclature

BOS	Background-Oriented Schlieren
CTA	Constant Temperature Anemometry
LDA	Laser Doppler Anemometry
PIV	Particle Image Velocimetry
POD	Proper Orthogonal Decomposition
VRS	Vortex Ring State
A_p	particle displaced fluid inertia force
B	particle Basset history force
C	tracer particle parameter
C_p	pressure coefficient
c	chord length; speed of light for Doppler effect
D	notional wind tunnel dimension; drag force
D_p	particle viscous drag force
d_p	particle diameter
$\underline{e}_b$	incident beam direction vector from laser

Supplementary Information The online version contains supplementary material available at https://doi.org/10.1007/978-3-031-12437-2_2.

R. Green (✉)
School of Engineering, University of Glasgow, Glasgow, Scotland
e-mail: richard.green@glasgow.ac.uk

$\underline{e}_{pr}$ Doppler shifted scattered light direction vector, particle to receiver
f frequency
f_b incident laser beam frequency
f_D frequency shift due to Doppler effect; difference in Doppler shifts for dual beam
f_r Doppler shifted light frequency perceived at receiver
N_s particle Stokes number, $\sqrt{\nu_f/\omega d_p^2}$
p fluid pressure
P force on particle due to flow field pressure gradient
R rotor radius
r radial ordinate
t time
U flow velocity
U_∞ wind tunnel free stream velocity
(u, v, w) fluid flow velocity components in Cartesian reference
u_f fluid velocity
$\underline{u}_p$ tracer particle velocity vector
u_p tracer particle velocity
u_{pr} particle radial velocity
u_θ fluid velocity in tangential direction
V relative velocity, flow speed ; hot-wire voltage
$\underline{V}$ fluid or particle velocity vector
(x, y, z) spatial ordinates in Cartesian sense

Greek Symbols

α aerodynamic angle of attack
α particle motion phase relative to fluid
α receiver axis direction for Doppler effect analysis
β aerodynamic sideslip angle; velocity direction for Doppler effect analysis
Δs particle displacement
$\Delta \underline{s}$ particle displacement vector
Δt time difference, PIV time delay
η particle to fluid velocity amplitude ratio
θ angle between incident beams
λ light frequency
ν_f fluid kinematic viscosity
ρ_f fluid density
ρ_p particle density
σ ratio of particle to fluid density
ξ parametric ordinate
ω circular frequency, angular velocity

2.1 Introduction

The flow around a rotorcraft in even its simplest mode of flight is extraordinarily complex. It comprises entanglements of multiple trailed vortex systems, regions where compressibility effects are significant, areas of separated flow, and the flow can be regarded as unsteady and subject to significant interaction effects. A successful experimental test campaign may have to isolate specific phenomena, or investigate performance aspects of the flight vehicle, but the complexity of the flow is always a challenge. Crucially, if the aeromechanics of the flight vehicle have to be represented, then test models can be very challenging to design. While issues with fixed wing wind tunnel testing might be limited by available test Reynolds and Mach numbers, testing for rotorcraft needs to consider additional parameters such as advance ratio, trim condition.

A good skill-set is required for a researcher to be able to conduct an experiment successfully. Not only is a thorough knowledge of fluid mechanics and aerodynamics a pre-requisite, but a solid appreciation of wider aspects of physics, optics, electronics are essential. For aircraft and related testing, the wind tunnel practitioner needs a wider appreciation of aeronautics, interest in aerodynamics is not sufficient of its own. Expertise with mathematics and statistics are expected, and computing systems for instrumentation and data analysis are tools of the trade. A wind tunnel control room in the middle of a test campaign should be a calm place, there is much planning and coordination required to bring a test programme to life, but plenty can go wrong, and the ability to think on one's feet, problem solve in the moment, get one's hands dirty are the types of activity to expect. The theoretician or CFD specialist without prior experience in experimental testing is encouraged to work on an experimental programme; the reader is referred to Ref. [1] for a wider understanding of theory and practice.

The reader is expected to have a clear idea of why an experimental test campaign should be done, but may not have any *a priori* familiarity with experimental testing. *All* experimental work requires extensive planning. Inexperience and excessive eagerness to get started are a bad combination that leads to much wasted time and resource. In this respect the chapter is an overview. It takes the reader through basic wind tunnel layout and operating limitations, provides a background to the most essential techniques that the reader needs to gain an awareness of, and finishes with an introduction to the more sophisticated optical flow field measurement methods. The experimentalist should not attempt these more advanced methods without first considering the capabilities of the apparently less sophisticated techniques, and a particular emphasis is placed upon the importance of doing force and moment measurement and on the researcher being productive.

The reader is encouraged to gain a thorough understanding of wind tunnels and test techniques before any experimental design and testing is attempted, and regular reference is made to some important reference publications.

As a final comment to this introduction, wind tunnels can be expensive to build and use. Operating economics are a concern. Their demise due to the development of CFD has been predicted for a long time, but recent initiatives such as the UK National Wind Tunnel Facility have recognised their strategic value as a research base.

2.2   The Wind Tunnel

There are facilities other than wind tunnels available for experimental aerodynamics (towing tanks, whirl tower, testing chamber, flight testing), but the wind tunnel is the most commonly used instrument. We restrict the discussion to low-speed tunnels, as this is appropriate for the flight regime of rotorcraft. The reader is referred to the text by Barlow, Rae and Pope [2] for a broad overview of low-speed wind tunnel testing.

The wind tunnel generates a conditioned flow at free stream speed U_∞ in a carefully controlled and monitored environment. Experiments are usually conducted in the working section, which are generally *closed jet* section or an *open jet section*. Figure 2.1 shows images of test models in an open section and closed section jet respectively. The open jet arrangement has no confining walls and has the advantage that the flow streamtube around the model is not so strictly confined as with the closed section. Access to the test model is relatively unrestricted, but the lack of a floor can lead to safety issues.

The flow may circulate around a loop (return), for which the advantage is lower running cost, or in the case of a non-return tunnel the flow is exhausted out into the atmosphere downstream of the test section. Thus, a wind tunnel may be described as closed-return, for example. The closed-return tunnel is a commonly found type,

Fig. 2.1 Wind tunnel working sections. **a** helicopter wind tunnel model (quarter scale Dauphin) in open-jet wind tunnel (VZLU, Prague, CZ); **b** 2-D dynamic stall model in closed section wind tunnel (HP tunnel, University of Glasgow, UK)

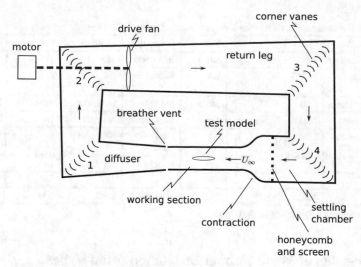

Fig. 2.2 Schematic diagram of a closed-return wind tunnel

and Fig. 2.2 shows a schematic diagram of a typical example. The flow direction in this diagram is clockwise.

The working section with test model are indicated. Atmospheric tunnels have breather vents at the end of the working section. Flow leaving the working section passes into a diverging diffuser and then encounters the first set of corner vanes. These turn the flow efficiently by 90° to a continued leg of the diffuser and another set of corner vanes 2. The flow then passes through the drive fan that raises the stagnation pressure of the flow. The return leg continues the diffusing section, there is corner vanes set 3, additional diffusion followed by corner vanes 4. The flow then passes into the settling chamber at its minimum speed, the function of the settling chamber is to give time for disturbances in the flow to settle and dampen out. Flow straightener and a screen are often fitted inside the settling chamber, which help to remove swirling motion and turbulence from the flow. The flow then passes through the contraction, where it is accelerated to the test section velocity U_∞. A general rule is that flow quality in the working section expressed in terms of turbulence level is improved with a higher ratio of contraction entry area to working section entry area (the contraction ratio). The disadvantage is a much larger wind tunnel size for a given working section size.

Wind tunnel operation costs are extremely important. The ideal wind tunnel is one that is as large as possible and runs at any desired speed. Very large wind tunnels occupy an enormous amount of space, power requirements increase as a function of the square of the size and the cube of the speed. Forces and moments on wind tunnel test models need consideration; achieving a very small increase of Reynolds number might be at the expense of a huge increase of dynamic pressure and model size, the model needs to be strong enough. A compromise has to be reached, and it is a case of understanding what type and size of wind tunnel is suited for a given task.

Table 2.1 Relative merits of instrumentation systems other than scientific benefits

Instrumentation	Cost	Complexity	Training	Calibration	Maintenance	Safety risk	Difficulty
Model positioning	VH	H	Some	Infrequent	Yes	H	L
Force/moment balance	M/VH	M	Some	Infrequent	No	L	L
Flow visualisation	L/M	M/H	Some	No	Little	M/H	M/H
Pressure probes	L/M	M	No	Infrequent	No	L	M
Pressure scanners	L/M	H	No	Generally no	No	L	M
CTA	M	H	Yes	Always	Wire repair	L	H
LDA	VH	VH	Yes	No	Yes	VH	VH
PIV	VH	VH	Yes	Always	Yes	VH	VH

L low, *M* moderate, *H* high, *VH* very high

2.3 Benefits of Wind Tunnel Instrumentation Types

A measurement technique should be used because of the scientific or engineering value of the data it provides. Instrumentation is costly, and some items are expensive to maintain and, in fact, difficult to live with. Furthermore, the user must bear in mind likely costs in the event of equipment failure. Skilful use of an inexpensive technique is a better achievement than poor use of one that relies on highly sophisticated instrumentation. This is important for the uninitiated, where the opportunity to learn is of paramount importance, and Table 2.1 is intended to give some idea of the pros and cons of some of the most widely available methods. As an example, the inconvenience of wire repair for the Constant Temperature Anemometry (CTA) technique is often cited as a reason to not use the hot-wire technique, but the wires themselves are often only broken during handling. The solution is a stock of replacements, and, ideally, a repair kit. Contrast this with the repair or replacement costs of a major system component for PIV, never mind the difficulties of working with class 4 lasers. Calibration for CTA is also cited as an inconvenience, but a calibration process for PIV must be also done, and neglect of the state of the elements of a PIV system that affect the calibration will result in an unknown drift and error. Calibration of instruments where it is necessary is an essential step in guaranteeing the quality of data, and should never be viewed as a hindrance.

2.4 Force and Moment Measurements

Wind tunnel test models need to be mounted in wind tunnels. This provides an obvious way of measuring the aerodynamic forces and moments, as the measurement system can be integrated with the model mounting and positioning system. *Most* wind tunnel testing is for force and moment measurement, this simple matter of fact can be quite easily forgotten. *Most* flow field investigation is only of interest because there is some effect on an aerodynamic force or moment to be investigated. The wind tunnel model shown in Fig. 2.1a is mounted on a stiff strut, which can be seen emerging below the belly of the model. The model fuselage is in turn attached to a six component force balance inside the model mounted at the end of the strut. The wind tunnel model shown has a separate loads measurement system for the stub rotor, allowing the hub drag to be separated from the fuselage drag. The wind tunnel model in Fig. 2.1b is instrumented with surface mounted pressure transducers (see later section), and pressure may be integrated to provide some information about the aerodynamic loading.

2.4.1 Model Positioning System and Balance Systems

The model must be positioned in the wind tunnel, and its attitude and orientation should be varied with ease. The simplest model positioning system should allow for variation of angle of attack, and more sophisticated turntable systems allow for variation of model yaw, pitch and roll angle. The flight dynamics requires aerodynamic data to be expressed in a particular way, and the aerodynamicist can frequently be ignorant of the sign conventions and flow directions that flight dynamicists use. Model attitude in terms of yaw, pitch, roll, for example, will need to be expressed in terms of angle of attack α and sideslip β.

Force measurement can be performed with something as simple as a single component load cell, but whatever device is used the researcher must be familiar with its design and theory of operation. Balances can be extremely delicate, but all balances are vulnerable to being mishandled. The model mounting position relative to the moment resolution centre of the balance must be known if sense is to be made of measured moments

External Force Balance. This type of balance is outside the wind tunnel; a photograph of such a balance is indicated in Fig. 2.3a. The model is mounted onto the balance by struts, and it allows for some yaw and roll angle adjustment in addition to pitch. The balance design is very stiff indeed, and the whole system benefits from high inertia in addition to the stiffness. This balance design produces particularly accurate forces and moments, and calibration can take place in-situ.

Sting Balance. This type of mounting and balance system offers the flexibility of high yaw, pitch and roll angles at the expense of mounting stiffness and smaller inertia. Ideally the sting and balance are inside the model, and this can make for a complex model installation procedure. A diagram of a sting-balance system is shown in Fig. 2.3b. A typical balance is shown in Fig. 2.4a, the flexural elements and strain gauges can be seen. The ground end fits to the sting, the model is mounted on the live end.

Rotating Shaft Balance. These are balances purpose built for mounting on rotating shafts. They can be used for propeller and rotor forces, for example for individual

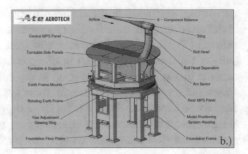

Fig. 2.3 External **a** and internal **b** wind tunnel balance systems. The external balance sits entirely underneath or over the working section, and the model is mounted to the balance via struts. (ATE Aerotech, UK)

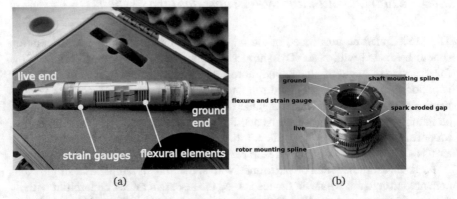

Fig. 2.4 Some six component balances. To the left, sting balance (ATE Aerotech, UK); to the right, rotating shaft balance (Aircraft Research Association, UK)

blade forces and moments. A rotating shaft balance is shown in Fig. 2.4b. It is manufactured from two concentric elements with an annular gap, the inner part is the ground side, connected to the rotor shaft, the outer part is the live side, connected to the rotor assembly. Flexural elements are manufactured that bridge the ground and live sides, and gaps between the live and ground side elsewhere are manufactured using a technique such as spark erosion. Excitation voltage signals to the strain gauges and strain gauge signals fed across are slip rings or via a telemetry system.

2.4.2 Balance Calibration

Balances have to be manufactured with great care, and handled in such a way. Careful balance calibration is essential, and this procedure must be documented. Specialised jigs are required, and calibration must take place in a carefully controlled environment. The calibration process only guarantees the best accuracy of the balance system. In use the model position will deform and balance signals drift, the user must be aware of this and factor these effects in when stating accuracy of the actual measurement.

2.5 Flow Visualisation

A flow visualisation test is regarded as an essential component of any test campaign. In some cases it may be completely impractical to do, but where it is possible, it should be attempted. Most flow visualisation methods are low cost, and a consumer camera is often all that is needed for image recording. Specialist or high-end cameras are only required where a particular performance demands it. Illumination methods include flood lights, UV lights, strobe lights and continuous or pulsed lasers. Safety precautions are often necessary; in the cases of use of lasers and stroboscopes this is obvious, but use of oil may create slip hazards. The flow visualisation method works by making some property of the flow visible, this may be to visualise the flow over the body surface, or to see features of the developing flow field. A flow field visualisation is recommended in advance of any test involving PIV. Some flow visualisation methods are as follows.

Surface Tufts. This is a useful technique for visualisation of attached or separated, steady or unsteady flows over a body. The tufts align with the local flow direction adjacent to the surface. Tufts are typically short pieces of thin thread stuck to the model surface with glue or tape. The tuft colour should be in good contrast with the surface, and fluorescent tufts illuminated with UV light create a very strong contrast, but normal lighting is the usual technique. A stills or video camera is sufficient for recording the tuft alignment. High speed video photography with appropriate image post-processing can show the fluctuation of the tuft position, useful for the interpretation of unsteady, separated flow.

China Clay. A mixture of china clay and paraffin is poured onto the model surface. The mixture must be runny enough for the flow to move the mixture, but thick enough so that it does not flow over vertical surfaces too easily. When the wind tunnel is turned on the wall shear stress moves the china clay mixture over the model surface, and the paraffin dries off. Surface streamlines and separation lines are revealed. It will show a time mean effect, and the mixture will tend to settle and remain wet in quiescent zones where surface shear is very low. China clay visualisation can create some beautiful images, and different coloured dye in the china clay can be used to indicate where specific areas of flow track to. It is useful for streamlined bodies, and is excellent for revealing the effects of excrescences and details at body junctions. The dried mixture is cleaned off easily after a test. Note that tape must be applied over any pressure tappings on the model, otherwise the tappings will be blocked.

Smoke Flow. Smoke flow patterns reveal details of the flow field; the user must be aware of the difference between a streakline, a streamline, and a particle path line. Smoke flow visualisation works best when the smoke plume is dense, flow field turbulence can mix the smoke plume reducing contrast. The injection point of the smoke into the flow is critical, it has to be entrained into the desired region of flow. Smoke plumes are typically created using a wand with a heated tip, oil is pumped onto the tip of the wand where it vapourises to form the distinct smoke. Moving the smoke wand around will show how the flow deviates over the various parts of a test body. Smoke flow can be used to visualise inflows, wakes, vortices, and can trace out separated zones. It is particularly useful for the visualisation of unsteady flows.

Schlieren and Shadowgraph. These techniques are usually associated with visualisation of transonic and supersonic flows, but they have been used to great effect for visualisation of flows around rotors and entire helicopters. Change of air density causes a change of refractive index, thus changing the path of a light ray. The Schlieren method reveals the spatial density gradient, shadowgraph the second derivative. The article by Bagai and Leishman [3] provides an analysis of Schlieren and shadowgraph methods for rotor flows, and the reader is encouraged to consult this article as a useful starting point. Ref. [4] shows an interesting application of the shadowgraph method. Recently, the technique of Background Oriented Schlieren (BOS) has been used to great effect. In this technique an image of a background pattern is compared with the image of the background pattern behind the flow field. The image processing required is lengthy, but the technique can be used in the laboratory and in flight testing, see the articles by Raffel [5] and Kindler et al. [6]. In particular, BOS has the advantage that some quite straightforward instrumentation is the only requirement.

2.6 Pressure Measurement

Pressure measurement in fluid mechanics is a particularly powerful technique. It is almost always used for determining the flow speed in the wind tunnel, and pressure can be measured to infer flow speed elsewhere or to determine surface flow features and be integrated to provide loads and moment information. Pressures measured on the wind tunnel wall surfaces can be used to calculate body lift, pressure measured in a body wake will provide drag, and these techniques can lead to low cost of model production. Pressure sensors tend to be in the form of pressure scanners or individual pressure transducers, and high performance pressure transducers have frequency responses up to several tens of kHz, so may be used for unsteady aerodynamics, measurement of turbulence phenomena, aeroacoustics. Pressure measurement by arrays of probes can provide all three velocity components at a point and even vorticity. An appreciation of pressure measurement is therefore a pre-requisite for any experimentalist.

Pitot-Static Probe. This type of pressure probe measure the flow stagnation pressure at the probe tip, and the probe static pressure at a location some distance behind the tip. The difference between these two pressures is the local dynamic pressure. How

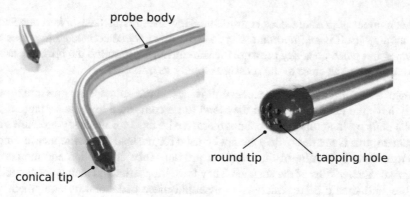

Fig. 2.5 Multi-hole probes for flow field pressure and velocity measurement (Surrey Sensors Ltd, UK)

the probe is used depends upon the application, but they are typically used to obtain wind tunnel free stream speed, and arrays of these probes are used for wake and jet surveys.

Multi-hole Probe. The basic Pitot-static probe cannot measure flow direction. Multi-hole probes consist of a specially profiled probe tip with a tapping hole at the front and an array of holes around the azimuth of the profiled tip, typically there are 5 or 7 holes in total. Calibrated correctly, they will provide data up to a flow angularity of 60° or so, and will provide local static pressure p, stagnation pressure p_o and (u, v, w) velocity. Pressure transducers placed close to the probe tip will give the probe a frequency response of 1kHz or so, and these types of probes can be used for unsteady flow field measurement. Images of multi-hole probes are shown in Fig. 2.5, these are mounted on a traverse and moved around the flow field. Probes such as these only require one comprehensive calibration that remains valid so long as the geometric properties of the probe are not altered; the reader is referred to Shaw-Ward et al. [7] for a description of a calibration process for this type of probe.

Wake or Jet Rake. A wake or jet rake is an array of fine diameter stagnation pressure probes mounted on a streamlined support. The probes are spaced at regular intervals, and the rake may be as large as practically allowable. They replace the job of a single probe mounted on a traverse, but it might be mounted on a traverse for additional wake resolution or for a spanwise traverse. Some of the probes on the rake are static pressure probes, and data from the rake are integrated using momentum theory.

Surface Pressure Tapping. Body surface pressures are measured using a pressure tapping. This is a small hole < 1mm diameter drilled into the model surface, this may then be connected to the pressure sensor. The model maker must be careful to ensure that the tapping is flush to the surface so that the tapping itself does not affect the local pressure. They must be at precise locations. There is always a compromise about the number of tappings that may be created, this may be due to space limitations or the number of pressure sensors available. How the pressure sensor is connected to the tapping itself is an important issue; a long tube may be connected to a pressure scanner, but when unsteady pressure measurements are required the pressure sensor must be located as close to the pressure tapping as possible.

Pressure Scanner. Pressure scanners allow measurement of large pressure arrays. They have the disadvantage that they tend to be connected to pressure tappings or probe ends by long tubes, and cannot therefore be used for unsteady pressure measurement; this is not to say they cannot be used for unsteady pressure measurement, the user must be careful of the length of pressure tube connection and the performance characteristics of the scanner. They have the distinct advantage of flexibility of use, and should be regarded as an essential piece of laboratory equipment that offer tremendous value for money if looked after and treated well. Modern scanners allow reconfiguration of subsets of the scanner array, so that multiple pressure ranges are available within the same module.

Pressure Transducers. Pressure transducers are the small sensors that may be used to convert the pressure into a signal for measurement. Individual pressure transducers are typically mounted close to surface pressure tappings for measurements of time-varying pressure at the model surface; they may be very small indeed and have a bandwidth of tens of kHz, but the lower cost type have limited bandwidth and tend to be larger in size. Pressure transducer cost has reduced over the last few decades, so that instrumenting a test model with surface mounted pressure transducers and leaving them in the model does not carry such a high financial penalty. Figure 2.6 showing a pressure transducer array mounted in the inside of the model, once the model is closed up the transducers are irretrievable. Note that the connection tube length from the surface pressure tapping to the transducer membrane is critical; pressure signals are attenuated and their phase changes when connected to a pressure pipe, if the tube is too long the attenuation and phase change become excessive, and the tube short be as short as possible.

Large arrays of pressure transducers have to use high channel count data acquisition systems that are able to exploit the bandwidth of each individual transducer. Some pressure transducers have only very small output voltages, amplifier arrays are necessary, so a suitable signal conditioning and data acquisition system has to be available. Pressure transducers mounted inside rotors or propellers will have their output signal fed out of the rotating environment using a slip ring array.

Pressure Sensitive Paint. Pressure sensitive paint (PSP) works on the principle of the partial pressure of oxygen in air and its effect on a paint substrate on the model surface. Images of the model surface during a test are recorded. The PSP method has the advantage that the model does not have to be instrumented with pressure tappings or pressure transducers, but it works best at higher dynamic pressures. It has promise for helicopter rotors, where the dynamic pressure is high.

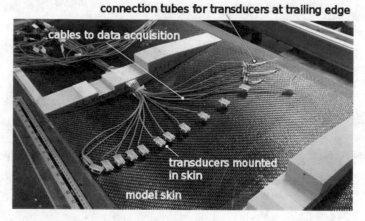

Fig. 2.6 Surface mounted pressure transducers mounted inside model (J. Walker, University of Glasgow, UK

2.7 Thermal Anemometry

Thermal anemometry includes the so-called hot-wire technique. It is an important method for flow velocity measurement, and it has the distinct advantage over all the other techniques of an exceptionally high bandwidth. Hot-wire probes are mounted on a traverse, and can be used for a wide range of flow fields if their operational requirements are understood. The technique is highly cost-effective in terms of the capital cost of the instrumentation, low maintenance costs, long shelf life of the parts, and should be present in any aeronautical laboratory. Thermal anemometry systems have various modes of operation, the one to be described further is that of constant temperature anemometry. The reader should refer to Ref. [1] and conduct thorough background reading on hot-wire theory and operation.

2.7.1 Theory of Thermal Anemometry

Figure 2.7 shows the basic principle of thermal anemometry; an electrical current heats a wire, convective heat transfer due to the flow speed U cools the wire down. A relationship between the flow speed and the electrical current must be established. In constant temperature anemometry, a control circuit keeps the wire at a constant temperature, and the wire voltage can be measured. The wire is typically very small diameter (a few microns) so that thermal inertia is low.

A wire is supported between prongs, the wire length might be 2mm or so. The flow speed U is at an arbitrary orientation to the wire, and the component of velocity driving the heat transfer is normal to the wire axis. Wires are arranged in pairs or as three to evaluate velocity components. A typical arrangement for a pair of wires is in the shape of an X (viewed from the side), where two slanted wires are arranged notionally at right angles with respect to one another.

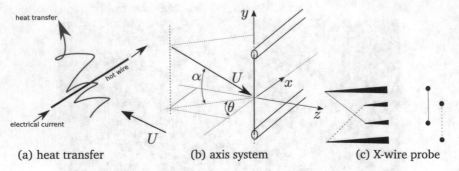

(a) heat transfer   (b) axis system   (c) X-wire probe

Fig. 2.7 Principle of thermal anemometry

2.7.2 Calibration

Calibration is an essential step for hot-wire anemometry. The established relationship between the wire voltage V and velocity U normal to the wire is through King's Law, $V^2 = A + BU^n$, where A, B, n are constants. Calibration is best done using a purpose built calibration rig that the wire is fitted into, this will generate a range of flow speeds and sample the wire voltage to evaluate the relationship. Cross-wires or triple wires are calibrated at known yaw and roll angles, so that the flow direction can be determined. Calibration does need to be done frequently, the wires get dirty for example.

2.7.3 Use of Hot Wires

The constant temperature anemometry technique may be used to evaluate the mean flow field, but it is particularly useful where turbulence quantities are required. The signal is continuous, and the measurement volume is small. Probes can be placed in difficult to access locations, and arrays of probes will provide data at many locations in the flow.

2.8 Flow Tracer Methods

Techniques for flow field measurement described so far all require the use of a probe. These have the disadvantage that the presence of the probe and its supporting apparatus may affect the flow field, and there is a wish to avoid this. Flow tracer methods rely on a tracer medium that follows the flow and may provide a signal that can be measured. They are most usually associated with flow field velocity measurement, This section will consider the dynamics of a tracer particle, and an introduction to particle image velocimetry and laser Doppler anemometry will be provided.

2.8.1 Tracer Particle Dynamics in 1-D Flows

The tracer particle is fundamental in providing the signal to be measured, it reflects or scatters incident light. An analysis of the tracer particle dynamics in a fluid flow is informative, as it has a fundamental impact on the choice of instrumentation and the limitations of the technique; the experimentalist should not take it for granted that the tracer particle follows the fluid flow exactly. The reader is referred to Rcf. [8] for a rigorous analysis of the dynamics of a flow tracer particle. We consider the motion of a small tracer particle of diameter d_p in 1-D flow of velocity u_f. In any tracer particle method, it is the particle velocity u_p that is measured. The particle does not simply follow the flow; the particle has forces exerted on it by the flow, and the particle accelerates in response to these forces. There is relative velocity $V = u_p - u_f$ between the particle and the fluid, and this is responsible for three of the forces acting on the particle as follows:

- Drag force D_p, this is the viscous drag force on the particle due to relative velocity V. For practical purposes the particle flow is a Stokes flow, for which a well known relationship exists;
- Force A_p required to accelerate fluid displaced by the particle. As the particle accelerates due to change of fluid velocity the mass of fluid displaced by the particle has to accelerate. It is a non-viscous force, and is easily modelled using potential flow theory;
- Basset history force B, which is a viscous force on the particle due to its unsteady motion. It is due to unsteady boundary layer development over the particle.

The remaining force P is due to pressure gradient in the fluid *in the direction of flow*. This is a streamwise buoyancy force. There are other effects on tracer particles due to pressure gradients in a flow, these will be discussed later. The equation of motion of the particle is then

$$\rho_p \frac{\pi d_p^3}{6} \frac{du_p}{dt} = -D_p - A_p - B + P. \tag{2.1}$$

Rcf. [8] provides a detailed discussion of this equation. We are interested in the motion of a particle intended to follow the flow speed, we are not interested in a two-phase flow for example, so we suppose that our tracer particle is small. The Bassett history force may be neglected for most applications. A consideration of the particle response to a sinusoidal fluid flow speed u_f at circular frequency ω is useful to consider, then a transfer function H is defined such that

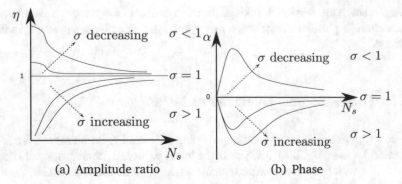

(a) Amplitude ratio   (b) Phase

Fig. 2.8 Amplitude ratio η and phase α of particle response in sinusoidally oscillating fluid flow speed u_f as a function of Stokes number N_s

$$u_p(i\omega) = H(i\omega)u_f(i\omega). \tag{2.2}$$

We then define an amplitude ratio $\eta = u_p/u_f$ and a phase angle α. Summary sketches of the variation of η and α as a function of Stokes number

$$N_s = \sqrt{\frac{v_f}{\omega d_p^2}} \tag{2.3}$$

are shown in Fig. 2.8, where curves for different particle: fluid density ratio $\sigma = \rho_p/\rho_f$ are shown.

This reveals some fundamentally important observations as follows:

- Amplitude ratio is 1 and phase angle is zero for $\sigma = 1$. This means the particle follows the flow perfectly;
- *Heavy particle* has $\sigma > 1$, the amplitude ratio is less than 1, the phase angle is negative. The particle speed lags the flow speed;
- *Light particle* has $\sigma < 1$, the amplitude ratio is greater than 1, the phase angle is positive. The particle speed leads the flow speed.

The optimal particle has $\sigma = 1$, it is neutrally buoyant. The change of behaviour either side of $\sigma = 1$ is due to the streamwise buoyancy. The choice of testing medium has a significant outcome on the availability of tracer particles as follows:

- Wind tunnel. Air density is typically $\rho_f = 1.225$ kgm^{-3}. It is extremely difficult to arrange for a neutrally buoyant particle; a gas filled bubble is possible, but is it practical?—Significant progress has been made with the use of helium filled soap bubbles, the technique is now viable. Furthermore, suitable particle substrates will be significantly denser than air. Tracer particles for air and gas flow measurement have $\sigma \gg 1$.
- Water tank. Water has density $\rho_f = 1000$ kgm^{-3}, there is a wide choice of particle substrate than can give $\sigma \simeq 1$.

Given that the requirement is that the tracer particle is required to follow the flow speed as closely as possible, experimentation in water as a fluid medium seems ideal. This is indeed the case, and water offers other benefits such as higher hydrodynamic forces, but water as a testing medium presents practical challenges that are difficult to overcome. We will restrict the discussion to problems appropriate for wind tunnels, but the experimentalist must be aware of the challenges specific to the testing medium.

Tracer Particle for Air Flow. We have established that we must accept a tracer particle density ratio $\sigma \gg 1$ for an air flow. Take Eq. 2.1 for the 1-D motion of the particle, an order of magnitude analysis then means that the particle Stokes drag D_p is the dominant term on the right hand side, so we solve

$$\rho_p \frac{\pi d_p^3}{6} \frac{du_p}{dt} = -3\pi \nu_f \rho_f d_p V. \tag{2.4}$$

Solution of Eq. 2.4 is straightforward, and we can consider two types of fluid flow as follows:

Sinusoidally Driven Fluid Flow. Consider a sinusoidally driven fluid flow speed u_f at circular frequency $\omega = 2\pi f$, we have a phase difference $\tan \alpha = \omega/C$ and particle: fluid amplitude ratio

$$\eta = \left(1 + \frac{\omega^2}{C^2}\right)^{-1/2}, \tag{2.5}$$

where $C = 18\nu_f/\sigma d_p^2$. A calculation for an air flow with $\sigma \approx 1000$, $d_p = 1\text{mm}$, $\nu_f = 1.5 \times 10^{-5}$ reveals negligible phase difference and very high amplitude ratio ($\eta > 0.99$) up to frequency around $f = 6$ kHz.

Step Change in Fluid Velocity. Solution is straightforward for a step change in fluid velocity from one constant value (e.g. $u_f = 0$) to another, e.g. $u_f = U$. We then have the solution

$$u_p(t) = U\left(1 - e^{-t/C}\right). \tag{2.6}$$

The particle speed approaches the fluid speed asymptotically. It is useful to consider $t_{0.99U}$, the time the particle takes to accelerate to $u_p = 0.99U$, then

$$t_{0.99U} = -C \log_e 0.01. \tag{2.7}$$

Fig. 2.9a shows physical particle response time for a range of density ratio σ up to $d_p = 100 \, \mu\text{m}$.

Due to the square relationship with d_p the response time becomes large as d_p becomes large. The question is, how quickly should a particle respond? A meaningful time scale is c/U, where c is, for example, a chord length, so a meaningful time response must be significantly smaller than this. The purpose of a tracer particle is to

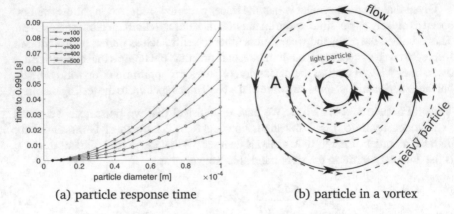

(a) particle response time   (b) particle in a vortex

Fig. 2.9 Tracer particle dynamics. **a** particle response time to reach 0.99 of step increase of fluid velocity U; **b** particle trajectory in a vortex flow, particle released from point A

provide the velocity information at that location, so the answer is that the response time should be as short as possible. A typical seeding substrate for a flow in air is oil with $\sigma = 660$, then a particle diameter of 1 micron gives $t_{0.99U}$ of about $11\mu s$. A 10 μm diameter particle has a response time $\sim 10^2$ times larger; this *may* be regarded as acceptable according to this analysis, but there are other issues to consider which will be discussed next.

Behaviour of a Tracer Particle in a Curved Flow Field. So far we have considered a 1-D flow only. A feature of a 2-D or 3-D flow is flow field curvature. Consider a 2-D vortex flow, we are interested in the tangential velocity u_0. There is a radial pressure gradient

$$\frac{\partial p}{\partial r} = -\rho_f \frac{u_\theta^2}{r} \tag{2.8}$$

that maintains the rotation, this is exactly sufficient to accelerate the fluid parcel such that it follows the curved path. Now consider the force required to make the particle follow the circular path around the vortex centre. The speed u_θ is constant, but the flow is accelerating everywhere in space due to the continuous change of direction. The step response calculation shown in Eq. 2.4 is not useful. A *fluid parcel* of notional diameter d_p requires force $\pi \rho_p d_p^3 u_\theta^2 / 6r$ to follow the circular path, and this is provided by the radial inward pressure gradient in the fluid. A *tracer particle* of notional diameter d_p requires force $\pi \rho_p d_p^3 u_\theta^2 / 6r$ to follow the circular path, but the radial inward pressure gradient in the fluid cannot provide this force in the case $\rho_p > \rho_f$, and the particle accelerates radially outward. Conversely in the case $\sigma \ll 1$ the force provided by the fluid radial pressure gradient exceeds the centripetal force required by the particle, and the particle accelerates radially inward. Figure 2.9b shows the trajectory of a heavy and a light particle in a vortex flow, the heavy particle moves outward as it orbits the vortex centre, the light particle moves inward as it orbits the vortex centre. In wind tunnel flows, this means that vortices are spotted easily by particle seeding voids at their centres. In addition to the uncertainty about velocity measurement accuracy, there is information loss where the particle seeding void exists.

To extend this analysis, we may suppose that the particle reaches a particle radial velocity u_{pr} in the vortex, then for $\sigma \gg 1$ the Stokes drag force in the radial direction balances the centripetal force for the particle, then

$$\frac{\pi \rho_f d_p^2}{6} \frac{u_\theta^2}{r} = 3\pi \nu_f \rho_f d_p u_{pr}, \tag{2.9}$$

Thus

$$u_{pr} = \frac{\sigma d_p^2}{18 \nu_f} \frac{u_\theta^2}{r}. \tag{2.10}$$

The terminal radial velocity u_{pr} is reduced with a very small particle diameter d_p. Consider an air flow with a vortex with $u_\theta = 20 \text{ms}^{-1}$ at $r = 0.05 \text{m}$. The 1 micron diameter particle with $\sigma = 660$ will have $u_{pr} = 0.02 \text{ms}^{-1}$. While there will be a seeding void due to higher acceleration at younger vortex age, this represents a negligible error in terms of velocity magnitude, but radial velocity measurements must be treated with caution.

Tracer Particle for a Wind Tunnel Flow. Much experience has been gained over decades of wind tunnel testing work using particle tracer methods. We have to accept that $\sigma \gg 1$, the particle will be heavy. To compensate for this so that the particle dynamics do not deviate excessively from the fluid dynamics, the tracer particle must be very small indeed. 1 micron is typical, with oil (e.g. olive oil) or water based solutions (polyethylene glycol) as the substrate. Even smaller particles include substances such as $Ti O_2$, where particle size is around 200nm, and these may be used in supersonic flows for example. Great caution must be taken with aerosol leakage, as sub-micron particles can enter the bloodstream directly through the lungs. The reader is encouraged to search the literature to gain a wider awareness of seeding tracer particles for wind tunnels, what is already available should not be taken for granted.

2.8.2 Optical Characteristics of Tracer Particles

We observe that a tracer particle in an air flow has to be extremely small, around 1 micron, for it to follow a fluid flow with an acceptably small error. By definition, methods using tracer particles rely then on reflection or scattering of light. Lorentz-Mie theory presents a solution of Maxwell's equations for the interaction of electromagnetic radiation with a sphere. Ref. [9] deals with this aspect of a flow tracer particle, and the reader is referred to this book and should read wider. The results

Fig. 2.10 Sketch of light
scattering power (dashed
circles indicate levels with
powers of 10) as a function
of direction for 1 micron and
10 micron particle size.
Sketch is approximate [9]

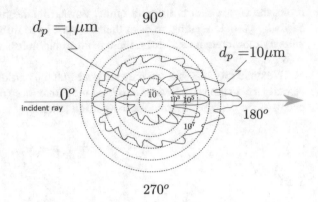

are key and have significant implications for the instrumentation requirements. A
small particle does not simply reflect light. We cannot simply consider scattering of
light by particles as a simple reflection problem. From a basic perspective light will
diffract around a small particle, some light will be reflected off the surface and the
light that passes into the particle will be refracted and suffer many internal reflections
until it then passes out of the particle. The number of internal reflections depends on
the Brewster angle, and this is the incident angle at which light will pass through a
refractive index interface. When viewed from a particular scattering angle, therefore,
a light ray may notionally have followed a number of beam paths, and the intensity
of the scattered light will vary strongly as a function of the scattering angle. Furthermore when the particle is sufficiently large constructive/destructive interference
of light waves will take place within the particle, and therefore the light intensity
will vary considerably with the scattering angle, with lobes of high intensity being a
distinctive feature of the scattered light. It is useful to consider thin film interference
as an example to help understand this.

A sketch of the Mie scattering profile for particles of 1 micron and 10 microns is
shown as Fig. 2.10. There are two key points from this:

- Scattered light power increases enormously with increase of particle size d_p;
- Scattered light power changes remarkably dependent upon scattering direction.
  Forward (180°) and backward (0°) scatter directions provide the highest power,
  while scattering to the side is orders of magnitude lower.

The latter point is demonstrated easily using a laser pointer; the beam is difficult
to see from the side, while it is easy to see viewing at a direction close to the axis
(point the laser way from your eyes!). The scattering behaviour explains the layout
of laser Doppler systems and drives the pulse energy requirements for lasers for
particle image velocimetry. Particle size effects mean that optical methods with tracer
particles are particularly well suited to flow measurement in water, as the demands
on instrumentation performance are less. All of this is driven by the need for flow
tracer particles to be small, and especially small for flow measurement in air where

$\sigma \gg 1$. Lasers tend to be used for specific experimental methods because they are capable of providing the intense illumination required and can be directed with great precision. Successful experimentation in water is possible using other light sources. Flow visualisation of smoke plumes can be done using floodlights or strobe lighting.

2.8.3 Particle Image Velocimetry

Particle image velocimetry (PIV) has emerged as one of the most important and powerful techniques for flow field measurement. Its operating principle is straightforward, and it provides the researcher with data that can provide a deep understanding of the behaviour of a fluid flow because it can provide velocity information over a wide part of the flow field at a near instant. It is especially well suited to unsteady, separated flows where transient phenomena occur. Its ubiquity in laboratories can act as a double-edged sword, however; it can be used in an inappropriate manner, for example measuring a flow using PIV becomes a goal of its own, and more informative techniques are forgotten about. The researcher needs to justify why PIV should be used, flow field information is only useful in support of other data. PIV for wind tunnels has the distinctive disadvantage that the instrumentation is extremely expensive. The inexperienced researcher will be misled by the apparent ease with which the commercially available PIV systems work, and it is a depressingly common occurrence that neophyte PhD students are allowed to walk into a laboratory knowing neither how to focus a camera nor calculate a PIV time delay. There is no substitute for understanding the technique, and the user manual of an off-the-shelf system is not the way to do this. PhD student supervisors who allow their students to enter a laboratory without thorough background knowledge and adequate training need to take note that they are exposing their students to great risk of personal harm due to the use of class 4 lasers, and the likelihood that very expensive instrumentation will be damaged beyond repair. In many respects an experimental campaign using PIV should be the *last* item on a list.

This section will describe the basic operating principle of PIV. The reader is referred to the book by Raffel et al. [9] and should read around the subject widely. In particular the reader should follow the development of PIV over the last few decades; what might now take a few seconds to compute in a laboratory will have taken many days of effort in the 1980s. The PIV practitioner needs a good understanding of optics and photography, be familiar with the theory of operation of lasers and their practical implementation, understand elements of signal and image processing, and have a fine dexterity and enormous reserves of patience.

Basic Principle of PIV. In its most basic form, PIV gives two component velocity field (u, v) in a plane (2D2C). Stereo PIV gives three velocity components (u, v, w) in a plane (2D3C). Both these methods infer the velocity from particle displacement in the measurement volume projected onto a recording medium, the measurement plane most usually defined by a very thin laser light sheet. Tomographic PIV will

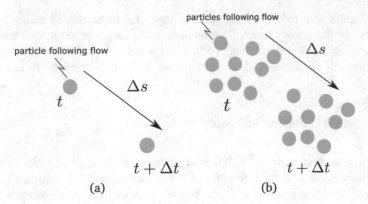

Fig. 2.11 Schematic diagram of basic operating principle of PIV: **a** basic principle and **b** movement of cluster of particles

give three velocity components (u, v, w) in a three dimensional volume (3D3C). *Never* attempt either stereo or tomographic PIV without first becoming competent in the use of 2D2C PIV.

Figure 2.11 shows a schematic diagram of the basic operating principle of PIV. The particle indicated at time t moves with the flow to the location at time $t + \Delta t$, the displacement is Δs. Velocity is defined as

$$\underline{V} = \lim_{\Delta t \to 0} \frac{\Delta \underline{s}}{\Delta t}, \tag{2.11}$$

Thus, PIV attempts to determine velocity directly and not by some other physical effect that can be measured. It needs to be understood that PIV provides velocity information over a finite but short time interval. The challenge with PIV is to determine the displacement vector $\Delta \underline{s}$. Δt is the PIV time delay; this needs to be short enough to satisfy the definition of velocity without excessive error, but long enough to determine particle displacement Δs with sufficient accuracy.

Ideally, each tracer particle in the flow is tracked individually, which would require each and every particle to be identified. In principle this is a good approach, but it is computationally awkward. Another matter is how particles are imaged. We know that a guideline particle size for a flow in air is 1 micron, optical wavelengths of light are then of the order of the particle size (a frequency-doubled Nd:YAG laser commonly used for PIV has a wavelength of 532 nm), so particle imaging is diffraction limited, and this is compounded by the spatial resolution of camera sensors. There is significant information loss about the particle. A practical approach to PIV is to seed the flow with many particles and use cross-correlation to track the movement of a pattern of particles, which has the advantage that individual particles do not have to be identified and the cross-correlation process is well understood. Figure 2.11b shows this notion. An image of a wide area of the flow field allows particle tracking or cross-correlation to be performed over the entire image by interrogation over small regions of the image, so that a picture of the flow field is built up.

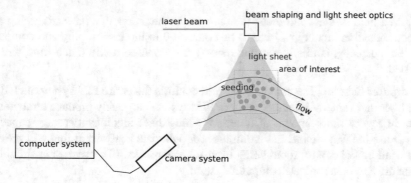

Fig. 2.12 Schematic diagram of the fundamental components of a PIV system arrangement

Instrumentation for PIV. This short section outlines what is required for a PIV experiment. It assumes the experiment has been set up and that optical access is available within the wind tunnel. The functioning of each component is critical, and the user must be familiar with the specification and functionality of each system component. A simple layout of a system set up for PIV is shown in Fig. 2.12.

Particle Seeding. This is the most fundamental element of the instrumentation. The seeder must be able to provide seeding for the wind tunnel at a sufficient rate, the larger the wind tunnel, the higher the seeding generation rate. Typical seeders use arrays of Laskin nozzles that atomise an oil substrate (e.g. olive oil), so a good compressed air source is required. Systems that heat mineral oil to create a mist that is pumped into the wind tunnel are also used. Oil can make a mess of the wind tunnel. DEHS can be used in these seeders, and alternatives include fog generators most normally used in the entertainment industry.

Illumination. For wind tunnels the most common illumination source is the pulsed Nd:YAG laser. The user must understand the laser arrangement and how the laser works, as a frequent source of problems is with the laser. The laser beam is delivered into the working section by a suitable arrangement of optics, a telescope to control the beam diameter, and the laser beam is expanded by a cylindrical lens into a thin sheet in the working section of the wind tunnel. All optical surfaces must be kept clean, all access windows must be blemish and scratch free. The user needs to know the flow area to be imaged beforehand, and be able to guide the light sheet into the correct position avoiding shadow as much as possible. This is a very difficult process that requires care and patience as class 4 lasers are usually involved. Tips and tricks include use of fluorescent tape and pens so that the laser can be seen through laser safety goggles.

Camera System. The size of the field of view that can be recorded successfully for PIV depends upon the laser power available and the camera resolution. These components need to be matched carefully. Cameras must be fitted with high quality lenses that can be focussed at the required stand-off distance and have a good enough f-number for image intensity. The user must be able to align the camera with the required field of view and focus the camera. Cameras for PIV have triggering and synchronisation inputs that have to be connected to the laser timing and computer system correctly, if this is not understood and done correctly much time will be wasted.

Computer System. The computer system controls the entire PIV system including the laser and camera timing. A modern PIV system can easily produce terabytes of data in a very short amount of time, so it must have significant storage capacity. *Never* use the wind tunnel PIV computer for bulk data analysis, as that will prevent the wind tunnel control room being used for other tasks. PIV processing should be done off-line on a separate computer system.

Spatial Calibration. Spatial calibration of the PIV system is an essential step. The camera used for PIV records the images as a pixel array, but the flow being measured is in physical space. Part of the experimental procedure is to record an image of a calibration grid at the same location as the PIV laser light sheet. The calibration grid is a known pattern, usually a pattern of dots on a contrasting background, and dot pitch is known precisely. A calibration grid of appropriate size is required. The calibration process provides the user with important information about the size of the field of view that can be recorded, and this information, along with knowledge of the flow speed and camera sensor resolution are required for calculation of the PIV time delay Δt.

Performance and Accuracy of PIV. A general rule for any experiment using any technique is to assume in the first instance that the results are not correct. They need to be checked carefully. The most common experience of researchers will tend to be through the use of a commercial PIV system, the mistake then is to assume it will do everything; they do not, and cannot in any way compensate for lack of knowledge of the user. PIV is very good at giving the user misleading data, and the user must be able to understand (1) the basic quality (fidelity) of the data from a PIV experiment and (2) the accuracy of a valid measurement. The only way to evaluate this is to have a thorough appreciation of the principle behind PIV, and to have done comprehensive background reading and, ideally, performed some numerical experiments using a PIV training utility. Data should be reviewed using a fresh pair of eyes, something obvious may emerge. At risk of provoking controversy, the accuracy of PIV can tend to be overstated. A realistic expectation is that PIV is a good tool for quantitative flow visualisation, and the most important aspect then is the fidelity of the result. The user needs to be familiar with:

- Random and systematic bias errors. A keen understanding of the instrumentation performance is the key to an appreciation of the latter. So-called peak locking in digital PIV is a bias error that all PIV users need an awareness of.
- Spatial resolution and over-sampling. How do the PIV analysis parameters affect the actual measurement resolution? Does over-sampling increase the measurement resolution?
- How do the vector validation methods employed work? What is their effect on the result?

The risk is that the user can overstate the capabilities of the system.

Post-processing of PIV Data. PIV provides a velocity field. What is subsequently done with those data depends upon the research goals. A common theme in this chapter is the advice that the researcher should understand the theory behind a technique, and for data post-processing this means an awareness of the mathematical demands for analysis of the type of discrete data that PIV provides. Simply using a function supplied with a system is not a good approach, as it may in fact be inappropriate. A good example of this is numerical differentiation for vorticity, for example; what is the best way of differentiating the data? A sequence of PIV images will allow mean flow and some statistics to be calculated; how many PIV images are required, and what do those statistics mean?—PIV provides snapshots of the flow, and this lends itself to a modal analysis, for example proper orthogonal decomposition (POD). A POD analysis will provide a set of modes whatever data are provided, but what do these modes mean?

Stereoscopic PIV. The basic 2D2C approach in PIV gives a notional (u, v) velocity in an (x, y) plane. Occasionally the three component velocity field (u, v, w) is required, where w is the velocity component normal to the measurement plane. This might be because the w component contains useful information, or perhaps the researcher is forced to use oblique viewing angles due to optical access issues. The particle displacement computed in PIV is from the projection of the particle movement in the volume of interest onto the recording medium. The reader is referred to the article by Prasad [10]. The projection of the recorded movement onto the focal plane coincident with the light sheet permits a relatively straightforward analysis of the perspective effect, and this is shown in Fig. 2.13. The most obvious camera view alignment is at a direction normal to the measurement plane, and perspective errors are negligible unless a wide area at short stand-off distance is being recorded. The difference in perceived particle displacement due to an oblique viewing angle is depicted in Fig. 2.13; the displacement perceived at oblique viewing angle θ is quite different from that of the normal view.

In stereo PIV, two PIV cameras are set at oblique viewing angles, usually at positive and negative θ relative to the normal, and the ideal angle between the two cameras is 90° or so. A rectangle viewed off axis becomes a trapezium, so there is significant distortion to be taken care of in the stereo processing methodology. Focussing the PIV cameras is a challenge, because the object plane is no longer normal to the lens principal axis so only a limited width of the light sheet comes into focus. A high lens f-number will increase the depth of field at the expense of less light entering the lens, but the usual approach is to use a so-called Scheimpflug mount that allows the lens and camera sensor axes to be tilted with respect to one another to give sharp focus across the field of view. Particularly careful focussing and calibration steps are required for stereo PIV, this allows the image distortion due to the off-axis view to be compensated for, and the stereo viewing angle for the stereo reconstruction may be computed. The process of setting up the instrumentation is much more difficult, and the data processing burden is more than twice that for two component PIV. It is

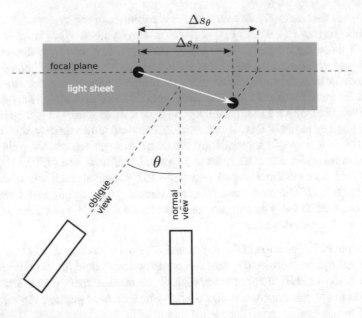

Fig. 2.13 Normal and oblique camera view at angle θ with PIV. The perceived particle displacements are Δs_n and Δs_θ respectively

for these reasons that the researcher should be very careful about the need to conduct stereo PIV in the first place, and only then to attempt it after sufficient experience with basic PIV has been gained.

2.8.4 Laser Doppler Anemometry

Doppler effect is a wave phenomenon in physics that is understood well. The laser Doppler anemometry method (LDA) came into being in the 1960s, as the development of the laser made an optical technique based upon Doppler effect a possibility. The reader is referred to Ref. [11] as a springboard for wider reading into the technique. As the technique relies on flow tracers, the requirements of the tracer particles are the over-riding concern (small diameter for gas flows). In turn this drives the specifications of the instrumentation.

Basic theory of LDA. Incident light illuminates the particle, and scattered light is picked up by a detector (the receiver). Doppler effect depends on the relative orientation of the incident light and the receiver, and the Doppler effect is invoked twice, firstly when the light hits the particle, and secondly when it is scattered to the receiver. It is essential that the reader understands the Doppler effect and how it is used in a practical LDA system.

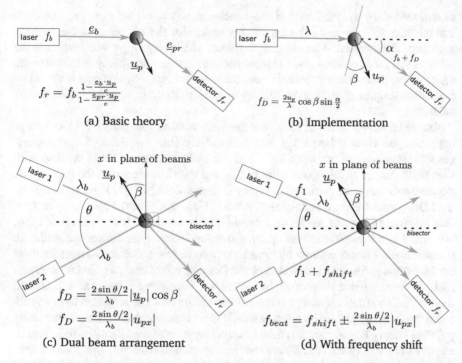

Fig. 2.14 Schematic diagram of transmitter (laser) and detector (receiver) with Doppler effect due to particle moving at velocity $\underline{u}_p$

Figure 2.14 shows an arrangement of a laser and receiver and how the Doppler effect theory is applied, with an illustration of the Doppler shift frequency and its dependency on the arrangement of the components and the direction of the particle motion. Note that in this sketch the receiver optic is arranged in a forward scatter direction to take advantage of the higher intensity of forward scattered light according to the Mie scattering theory. A typical laser for LDA has wavelength around 532 nm, therefore frequency $f_b = 5.6 \times 10^{14}$ Hz. Doppler frequency shift is typically less than 4 MHz per unit speed u_p, so the Doppler shift is less than one part in 10^8. This is tremendously difficult to measure, and a common implementation is to use a second laser beam with different incident direction and hence different Doppler shift. This is shown in Fig. 2.14 graphs (c) and (d), where the two beams coincide at a point. There is different Doppler shift frequency due to the different directions of the incident beams relative to the receiver, and the difference between the two Doppler shifts is indicated in graph (c). Note that the frequency difference does not depend upon the angle of the detector relative to the laser sources, only to the bisector angle θ. The sense of velocity detected is defined by the plane of the two beams, and the velocity component sensed by this arrangement is u_{px} in the direction x in the plane of the two beams normal to the bisector line between the two beams. The difference in Doppler shifted frequency f_D manifests at the detector as an optical heterodyne, a beating, and this can be measured. The problem remains that the direction of the velocity u_{px} cannot be resolved. If one laser is frequency shifted by f_{shift} with respect to the other, as in frame (d), then given that the detector is still individually sensitive to the relative directions of the incident beams, the beating frequency detected is biased by the shifted frequency, so that the beating frequency detected reveals the direction of velocity.

The reader is encouraged to follow the mathematics. The detected velocity u_{px} depends upon the beat frequency f_{beat} detected, the shift frequency f_{shift} imposed, the wavelength λ_b of the laser light, and the beam separation angle θ. Thus with high performance electro-optic components and with knowledge of the angle θ, the device needs no calibration, and accuracy is fundamentally high. If set up correctly, the LDA output will be unambiguous. A convenient model of the frequency shifted, dual beam LDA that the reader may consider is the so-called fringe model of LDA, where the tracer particle moves across a moving interference fringe pattern in the measurement volume defined by where the two beams coincide. A mathematical model of this gives the same result as the Doppler effect analysis, and it is almost as if the method is a time-of-flight technique. Whichever is used as a preferred description, the fundamental physical effect that the detector responds to is a time-changing intensity caused by interference. If there is any optical surface in any beam path that affects the polarisation state of the incident or scattered light, the interference effect will be extinguished; LDA cannot be expected to work reliably, for example, if there are any laminated or toughened glass, perspex or polycarbonate windows in the beam path. This is why a fundamental understanding of the LDA technique and its implementation are essential.

Typical laboratory implementation of LDA. Laboratory LDA systems take on many guises. The user has the freedom to tailor the system set up to match an experiment. The fundamental components are the lasers, the laser beam probe delivery heads, detectors and analysers and traverse system.

- Class-4 lasers are typically used;
- A typical modern laboratory will have the lasers fed into the optical heads by optical fibre;
- Some arrangements have the detector in the same probe unit as the laser delivery head, this exploits back scatter and makes for a very compact arrangement. Otherwise the detector might be in a different module;
- A two-component probe will take two sets of lasers. The beam pairs are arranged at right angles to allow the probe to measure two components of velocity. The actual physical velocity measured relative to wind tunnel axes then depends upon the direction of the probe axis, and that is the user's choice;
- Three components of velocity at a point may be measured by arranging three single component LDA probes or one dual component and one single component probe and bringing them to focus at a precise point. The angles of the probe axes relative to one another need to be known;

- Tracer particles arrive in the measurement volume at irregular intervals, as a particle passes through the measurement volume it emits a burst, and the burst spectrum analyser processes the signal and determines the LDA velocity during the burst.

Running an LDA Test. A wind tunnel and other experimental systems need to be run for a long duration in an LDA test. The researcher needs to be aware of a few key points.

- Do not be too ambitious with the number of measurement points early on in a test campaign. Higher spatial resolution scans can target areas of interest later on.
- Ensure the LDA traverse has the freedom to move over the planned scanning volume.
- Anticipate where the LDA beams may be blocked, and where strong reflections may occur.
- Do not allow LDA burst (count) rates to drop to too low a level, wind tunnel or test chamber seeding will have to be topped up.
- For dual beam LDA, check laser power levels are well-balanced while setting up the system. Check the balance as frequently as possible.
- Monitor the LDA performance during a test.

2.9 Case Studies and Results

We conclude this chapter with a short selection of experimental studies involving helicopter rotors and in some cases the complete vehicle.

2.9.1 Forces and Moments on a Model Helicopter Near an Obstacle

Helicopters frequently operate close to the ground and to buildings, the flow around the rotor is influenced by the presence of both. Figure 2.15 shows a helicopter model in a wind tunnel close to an obstacle on the ground [12]. The rig is in a large wind tunnel, and the helicopter model has a force balance on the end of the sting. The helicopter model contains a six axis load cell. The model was moved around using the traverse system supporting the model. The orientation of the helicopter model relative to the obstacle has a significant effect on the rotor forces and moments. The thrust data are plotted on the right; Z is vertical height of the model above ground. The test cases are T3 rotor in symmetry plane of obstacle and T4 a rotor facing an obstacle edge.

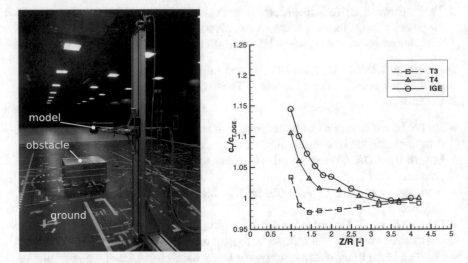

Fig. 2.15 Helicopter model close to an obstacle [12]. (D Zagaglia, Politecnico Milano, IT, 2016)

2.9.2 Dynamic Stall: Surface-mounted Pressure Transducers

Wind tunnel test campaigns for dynamic stall include experiments with models instrumented with surface mounted pressure transducers. Figure 2.1b shows such a test model. Sample data are shown in Fig. 2.16, from Ref. [13], where the development of the surface pressure distribution over the pitching cycle and the quarter chord pitching moment are shown. Surface pressure data show the build up of leading edge suction, its collapse and generation of the dynamic stall suction ridge, and the re-establishment of the fully attached flow. Pitching moment data are from integration of the pressure data.

2.9.3 Inflow to a Rotor: Flow Measurement with LDA

LDA measurements of a hovering rotor were performed, rotor diameter 1m. LDA was used to scan the inflow at a distance 1cm above the rotor disc. This was part of a study on the behaviour of rotor systems and rotor wakes in the vicinity of a neighbouring obstacle. Figure 2.17 shows the inflow velocity, positive into the disc [14]. Measurement of the hover inflow was compared with the flow in the presence of an obstacle beneath a portion of the rotor disc, and a load cell was used to determine thrust and trim state (Data from GARTEUR AG22 [15]).

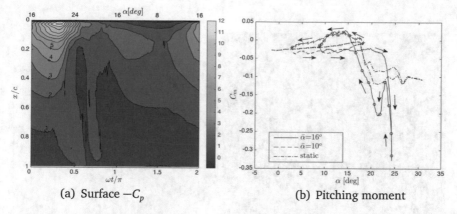

(a) Surface $-C_p$

(b) Pitching moment

Fig. 2.16 Dynamic stall measured using surface mounted pressure transducers. Sinusoidal pitch oscillation at reduced frequency 0.103, pitch amplitude 8°, mean angle $\bar{\alpha} = 16^o$, Re $= 1.5 \cdot 10^6$ (Y. Wang, University of Glasgow, UK, 2007 [13])

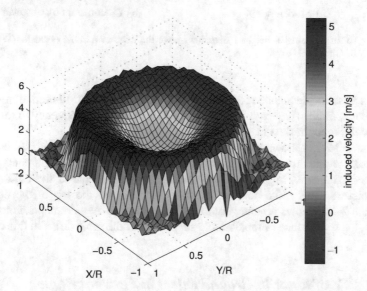

Fig. 2.17 LDA measurement of inflow induced by a hovering rotor; positive sense into the disc (D Zagaglia, University of Glasgow [14])

2.9.4 Smoke Flow Visualisation and Pressure Measurement

Smoke flow of the wake of a rotor hovering above a cuboid obstacle were performed [16]. If the smoke wand is placed close enough to the edge of the rotor disc, the smoke is entrained around the trailed vortex filaments. These filaments are persistent, and the smoke will reveal the helical vortex wake for several revolutions before the smoke is diffused or the wake breaks down. Figure 2.18 shows an image from a

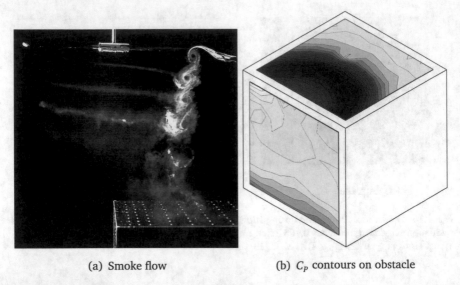

(a) Smoke flow   (b) C_P contours on obstacle

Fig. 2.18 Rotor hovering at height 1 diameter above the edge of a cubic obstacle. (D Pickles, University of Glasgow [16])

flow visualisation sequence recorded by a high speed camera. Illumination was by stroboscope, strobe frequency was slightly different to rotor frequency to reveal the flow development in time in a strobed sense.

The graph 2.18a shows flow visualisation using a smoke wand with stroboscope. The smoke wraps around the vortex cores, and the trailed vortex filaments can be seen, these filaments impinge on the surface of the obstacle at the bottom right of the picture. The white dots are pressure tappings on the top of the obstacle. The graph 2.18b shows pressure contours from pressure messurements, colours are positive, the top surface of the obstacle supports half the rotor thrust in this case.

2.9.5 PIV of Rotor in Ground Effect in Forward Flight

A 1-metre diameter rotor system [17] was sting mounted in a wind tunnel over a rolling ground, ground speed matched wind tunnel speed. The rotor system was mounted on a balance, and cyclic inputs allowed the in-plane moments to be trimmed out. Stereo PIV was performed over a 1m wide area, seeding is olive oil, $1\,\mu$m particle size, relative density $\sigma = 800$. Data are plotted in Fig. 2.19, which shows an instantaneous vector field. The centre of is rotor at coordinates $(0, 0)$, the ground is at $y/R = -1$. The noisy w contours for $x/R > 2$ are due to edge of light sheet effects.

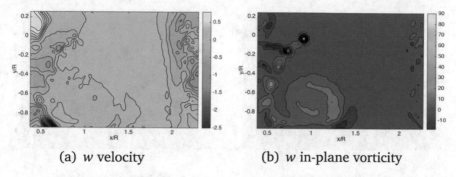

(a) w velocity   (b) w in-plane vorticity

Fig. 2.19 Stereo PIV of a rotor in forward flight ground effect; normalised advance ratio 0.66. 1 m wide field of view, wind tunnel flow is from right to left. (N. Nathan, University of Glasgow; data from GARTEUR, Ref. [17]

Much effort was made with the stereo PIV, but the w-component data were of limited value. Noisy w data on the right hand side of the graph 2.19a are from edge of light sheet effects (low intensity). So errors are larger. Vorticity in graph 2.19b shows strong positive vorticity of the trailed vortex filaments, the roll up of these filaments into the distinctive ground vortex, and negative vorticity is from the separation induced at the ground by the ground vortex.

2.9.6 PIV of Rotor Vortex Ring State

These figures demonstrate the remarkable breakdown of the usually expected helical trailed vortex wake into the *vortex ring state* (VRS) under certain conditions of descent. PIV measurements were taken of the flow around a rotor towed along a long towing tank to simulate the descent [18]. Seeding material used was silver coated microspheres, particle diameter $100\,\mu$m, relative density $\sigma = 0.9$. Rotor thrust was measured simultaneously so that flow field state could be correlated with thrust phase. Figure 2.20 shows out-of-plane vorticity and instantaneous streamlines around low thrust and high thrust.

The experiments have been carried out in water with time-resolved PIV and simultaneous thrust measurement. The leading edge of rotor disc is shown. The rotor centre is set at coordinates $(0, 0)$. The positive vorticity trailed from the rotor tip and the negative trailed from rotor root. The VRS is characterised by the formation and shedding of a toroidal vortex that contains the entire rotor. This vortex system is formed from vortex system trailed from the rotor tip.

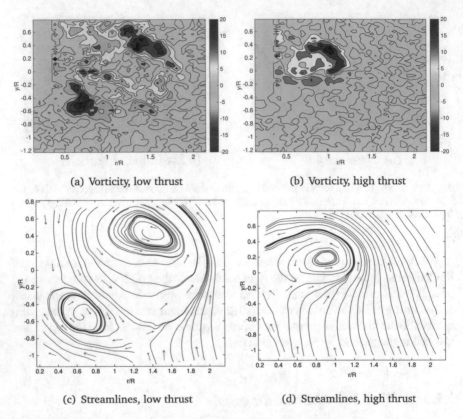

(a) Vorticity, low thrust   (b) Vorticity, high thrust

(c) Streamlines, low thrust   (d) Streamlines, high thrust

Fig. 2.20 PIV of vortex ring state of a rotor in off-axis descent, vorticity and streamlines in symmetry plane (Ö Savaş, University of California at Berkeley [18])

References

1. Goldstein RJ (ed) (1996) Fluid mechanics measurements, 3nd ed. Taylor & Francis. https://doi.org/10.1201/9780203755723
2. Barlow JB, Rae WH, Pope A (1999) Low-speed wind tunnel testing. Wiley-Interscience
3. Bagai A, Leishman JG (1993) Flow visualization of compressible vortex structures using density gradient techniques. Exper Fluids 15:431–442. https://doi.org/10.1007/BF00191786
4. Leishman GJ, Bhagwat MJ, Ananthan S (2004) The vortex ring state as a spatially and temporally developing wake instability. J Am Helicopter Soc 49(2):160–175
5. Raffel M (2015) Background-oriented Schlieren techniques. Exper Fluids 56. https://doi.org/10.1007/s00348-015-1927-5
6. Kindler K, Goldhahn E, Leopold F, Raffel M (2007) Recent developments in background oriented Schlieren methods for rotor blade tip vortex measurements. Exper Fluids 43:233–240. https://doi.org/10.1007/s00348-007-0328-9
7. Shaw-Ward S, Titchmarsh A, Birch DM (2015) Calibration and use of n-hole velocity probes. AIAA J 53. https://doi.org/10.2514/1.J053130
8. Emrich RJ (ed) (1981) Methods of experimental physics, vol 18. Academic Press, Fluid Dynamics, Parts A and B

9. Raffel M, Willert CE, Scarano F, Kahler C, Wereley ST, Kompenhans J (2018) Particle image velocimetry. A Practical Guide Springer. https://doi.org/10.1007/978-3-319-68852-7
10. Prasad AK (2000) Stereoscopic particle image velocimetry. Exper Fluids 29:103–116. https://doi.org/10.1007/s003480000143
11. Albrecht HE, Borys M, Damaschke N, Tropea C (2003) Laser doppler and phase doppler measurement techniques. Springer
12. Zagaglia D, Zanotti A, Gibertini G (2018) Analysis of the loads acting on the rotor of a helicopter model close to an obstacle in moderate windy conditions. Aerospace Sci Technol 78:580–592. https://doi.org/10.1016/j.ast.2018.05.019
13. Prince SA, Green RB, Coton FN, Wang Y (2019) The effect of steady and pulsed air jet vortex generator blowing on an airfoil section model undergoing sinusoidal pitching. J Am Helicopter Soc 64:032004. https://doi.org/10.4050/JAHS.64.032004
14. Zagaglia D, Giuni M, and Green RB (2018) Investigation of the rotor-obstacle aerodynamic interaction in hovering flight. J Am Helicopter Soc 63:032007–1– 032007–12. https://doi.org/10.4050/JAHS.63.032007
15. Visingardi A, De Gregorio F, Schwarz T, Schmid M, Bakker R, Voutsinas S, Gallas Q, Boissard R, Gibertini G, Zagaglia D, Barakos G, Green R, Chirico G, and Giuni M (2017) Forces on obstacles in rotor wake—A GARTEUR action group. 43rd European Rotorcraft Forum, 12th-15th September
16. Pickles D, Green RB, Giuni M (2018) Rotor wake interactions with an obstacle on the ground. Aeronaut J 122(1251):798–820. https://doi.org/10.1017/aer.2018.7
17. Nathan ND, Green RB (2012) The flow around a model helicopter main rotor in ground effect. Exp Fluids 52(1):151–166. https://doi.org/10.1007/s00348-011-1212-1
18. Savaş Ö, Green RB, and Caradonna FX (2009) Coupled thrust and vorticity dynamics during vortex ring state. J Am Helicopter Soc 54:022001–1–0022001–10, 2009. https://doi.org/10.4050/JAHS.54.022001

Richard Green is at the School of Engineering, Glasgow University, where he specialises in wind tunnel testing and instrumentation, a wide array of aerodynamic flows, particularly for rotor systems.

Chapter 3
Rotorcraft Propulsion Systems

Antonio Filippone and Nicholas Bojdo

Abstract This chapter introduces rotorcraft engines. the discussion is largely limited to gas turbine engines that dominate the field of modern helicopters. The Chapter is split into four main parts: (1) rotorcraft power plants and power trains; (2) engine ratings (with certification requirements); (3) performance envelopes; (4) intake protection systems. Rotorcraft engines are seldom treated as part of core rotorcraft engineering, since they are considered an element of propulsion; thus, they are associated to a different discipline. In this Chapter we demonstrate that there are peculiarities in this type of engines. Their integration into the airframe via transmission systems, rotor head, intake separators is unlike any fixed-wing vehicle.

Nomenculture

EGT	Exhaust gas temperature
IBF	Inlet barrier filter
IPS	Inertial particle separator
LTO	Landing and take-off
MCP	Maximum continuous power
MEP	Maximum emergency power
MTOP	Maximum take-off power
nvPM	Non-volatile particulate matter
OAT	Outside air temperature
TAS	True air speed (called V in equations)
SFC	Specific fuel consumption
UHC	Uncombusted hydrocarbons
VTS	Vortex tubes separator

A. Filippone (✉) · N. Bojdo
School of Engineering, The University of Manchester, Manchester, UK
e-mail: a.filippone@manchester.ac.uk

© The Author(s), under exclusive license to Springer Nature Switzerland AG 2023
A. Filippone and G. Barakos (eds.), *Lecture Notes in Rotorcraft Engineering*,
Springer Aerospace Technology, https://doi.org/10.1007/978-3-031-12437-2_3

EI	Emission index
A_e, B_e	Coefficients in Eq. 3.3
c_1, c_2	Empirical factors in pressure loss Eq. 3.7
$C_{1,c}$	Dust concentration
d_p	Particle diameter
k_{loss}	Power loss in a gearbox
k_1, k_2	Lower loss factors
M	True Mach number
n	Number of operating engines
N_b	Number of blades or vanes per stage
N_{gg}	Gas generator turbine speed, rpm
N_p	Power turbine speed, rpm
P	Engine power, kW
P_s	Engine power, SHP
$\mathscr{R}$	Gearbox reduction ratio
t	Time, s
T, T_1	Air temperatures, K
T_3	Total gas temperature at compressor exit, K
T_4	Total gas temperature at combustor, K
W	Gross or take-off weight, kg
W_1	Air flow rate, m^3/s
W_{f6}	Fuel flow per engine, kg/s
z	Flight altitude (m or feet)

Greek Symbols

δ	Relative air pressure
η_s	Particle separation efficiency
θ	Relative air temperature
Π	Overall pressure ratio
σ	Relative air density
τ	Time constant of engine deterioration
ω	Total pressure loss coefficient
$\overline{(.)}$	Mean value
$(.)_i$	Index counter
$(.)_{AP}$	Engine in final approach
$(.)_{ID}$	Engine in idle
$(.)_{TO}$	Engine in take-off
$(.)_{sl}$	Sea level

3.1   Rotorcraft Power Plants

The great majority of helicopter power plants is made of gas turbine engines tuned for delivery of torque. These engines are called *turboshaft*. They have become popular in the 1960s, with rapid advances in gas turbine engine technology. Prior to that time, helicopters were powered by Internal Combustion (IC) engines. IC engines are nowadays limited to historic aircraft (for example, the *Sikorsky S-55* and the *Kamov Ka-26*) and to light modern aircraft (for example, the *Robinson R-22* and the *Schweitzer S300*). In recent years there has been research to reinvigorate the use of IC engines on rotorcraft, with initiatives that addressed the development of diesel engines. The outlook for the future is possibly different, with new types of rotorcraft being proposed, some sub-scale, both manned and unmanned, that are to be powered by electrical machines or by hybrid systems. Electrical tail rotors have been proposed. The technology is in rapid development, but for the time being gas turbine engines are the only realistic options for full scale, heavy lift and military rotorcraft.

The first gas turbine helicopters were the French Sud Aérospatiale *Alouette II* in 1956–1957, powered by a Turbomeca *Artouste*, and the Bell UH-1A *Iroquois* powered by a Lycoming T53-L-1 engine, around the same time (the exact dates are a matter of dispute). The now historical Sikorsky S-55 is a particularly interesting example of IC engine integration into a rotorcraft. Figure 3.1b shows a photo of one such vehicle. The engine exhaust is a large pipe at the front of the airframe pointing right (from the pilot's view). The engine itself was mounted at the front of the aircraft, and the main shaft runs through the cabin to engage the main gearbox on top of the fuselage, Fig. 3.1b. Getting into one pilot's seat may require ducking under the main shaft to climb out onto the other side.

One of the key limiters of the IC engines is the relatively low power/weight, higher specific fuel consumption and large dimensions. Modern turboshaft engines offer very high power/ratio performance are extremely compact and can be coupled in twin- and three-engine configurations. An example of turboshaft cut-out diagram is shown in Fig. 3.2.

Rotorcraft engine configurations are very compact. The lack of space available on the aircraft has forced engineers to find ever more ingenious solutions to integrate the engine onto the airframe and the rotor system. Twin-engine configurations are normally mounted side-by-side. Three-engine configurations generally have two engines mounted side-by-side and one central engine, often mounted higher and aft (most Mil helicopters, the CH-53, the AW101, and their variants). The intakes are generally front-facing, though not always, and the exhausts are forced either backwards or sideways. Upwards exhausts would interfere with the rotor and would damage the blades [1]; downwards exhausts are an operational hazard and would prevent many operations requiring engines running.

Important elements in the engine-airframe integration include the full transmission system, the fuel systems and the engine cooling system (lubrication pumps, oil and afferent lines). We will describe briefly some of these systems.

Fig. 3.1 Author's photo (circa 1997) and diagram of the Sikorsky S-55 transmission system. Main shaft running through the cabin

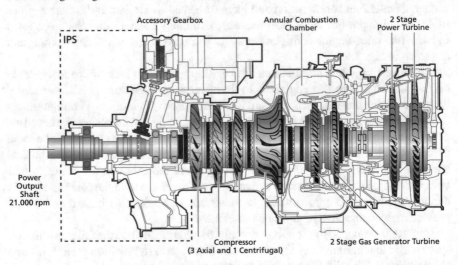

Fig. 3.2 Diagram of the the RTM322 turboshaft engine, adapted from a marketing brochure

Modern helicopter engines have a full electronic control system called FADEC, which is a hardware-software kit that allows full control and health monitoring of the engine. Several sensors in critical sections of the engine include temperature gages, pressure gages, and accelerometers. Add-ons include the intake protection systems, which are discussed in Sect. 3.5.

3.1.1 Architecture of a Turboshaft Engine

Torque-delivering gas turbine engines are classified as turboprops and turboshafts. The main difference between the two configurations is that the turboprop is suited to fixed-wing aircraft and delivers some residual thrust through the exhausts. This residual thrust makes up a non-negligible fraction of the net thrust, up to 10% or more at full throttle on the ground. The turboshaft is designed specifically for rotorcraft applications and does not provide any meaningful thrust in the exhaust. The key aspect of this engine is the ability to deliver as much torque as possible, which is then transferred to the rotor system.

Turboshaft engines have typically two main shafts. A gas-generator shaft consists of a gas turbine with a multi-stage compressor. The gas turbine can be a single- or multiple stages, depending on the engine rating. The multi-stage compressor has a number of axial compressor stages followed by a single-stage centrifugal compressor (or impeller). The reason why there is such a complicated turbomachinery is that centrifugal compressors deliver high compression ratios in a single stage and therefore are compact, whilst boosting the gas pressure to sufficient levels before it enters the combustion chamber. This group of turbomachinery rotates at very high rpm, and the design speed is called N_{gg} ('gg' is for gas generator). This rpm is often given in percent of the design rpm.

For reasons unknown to this author, overall compression ratios of turboshaft engines are seldom advertised, unlike turbofan engines. In any case, a modern single-stage axial compressor would have a compression ratio ~ 1.3, and a centrifugal impeller would have a compression ratio ~ 6.5. Therefore, the engine shown in Fig. 3.2 would be expected to have an overall compression ratio of $\Pi = 14.2–14.5$, unless otherwise specified by the manufacturer.

A co-axial shaft couples another gas axial turbine (single or multi-stage) with the output shaft, which then delivers power to the main rotor via a series of gear boxes. The angular speed of this shaft is virtually constant, but there are margins of short-term variation, for example within the range 95–110%. This angular speed is called N_p ('p' is for power turbine). If the torque on the output shaft increases, so it does too on the coupled gas turbine. The increase in torque would tend to slow down the turbine, that that can be prevented by increasing the fuel flow, which increases the mass flow rate of the hot gases through the turbine until the torque is restored. *Helicopter rotor speeds are constant, regardless of power output, with the exception of sharp increase/decrease in torque.*

An estimate of the power losses in the gear-box is done with the following assumptions. If the reduction-ratio is $\mathcal{R} = Q_{out}/Q_{in}$, and Q denotes the torque

$$k_{loss} = \frac{P_{in}}{P_{out}} = \frac{1}{\mathcal{R}} \left(\frac{Q_{in}}{Q_{out}} \right) = f(Q_{in}) \simeq k_1 \left(\frac{Q_{in}}{Q_{design}} \right)^2 + k_2 \qquad (3.1)$$

where k_{loss}, k_1 and k_2 are loss factors. Equation 3.1 implies that there is a floor in the power losses and that these losses increase with the input torque, which is normalised with the design torque. The term k_2 takes into account the fact that for small values of the torque, no output is expected, due to internal losses in the gear-box; the factor k_1 is to be determined experimentally, as it depends on the specific gear-box geometry.

3.2 Transmission and Power Train

The power transmission diagram of a modern helicopter is shown in Fig. 3.3. There is a gear-box to reduce the high speed of the engine shaft to a relatively low speed of the main rotor (reduction of 1:20 is not unusual). There is also a direct coupling of the main and tail rotor, via the gear-box and a central shaft, so that the two rotors rpm are locked.

The twin-engine configuration has a peculiar design. The availability of two engines would by itself not be a major engineering breakthrough if it were not for the direct coupling of two engine shafts to a main power train. One example of this architecture is shown in Fig. 3.3 for the AS-332 *Super Puma*, a helicopter that often operates in challenging environments over open occan. There is a left and right engine, with shaft outputs at very high speed, $N_p = 22{,}841$ rpm. These shafts engage a main central/horizontal shaft through a series of gears and free-wheels that reduce the rpm to 4,888 rpm (reduction ratio approximately 1:4.5). There is a main gear box just ahead of the central shaft that through a series of epicyclical gears delivers torque at 265 rpm (rotor design speed). The tail rotor is coupled with the central shaft through a series of gear boxes: an intermediate gear box changes the direction of the transmission and the final tail rotor gear box delivers power at 1279 rpm. The ratio between tail- and main rotor speeds is a fixed ratio of 4.826.

Should one engine shut down because of damage, fire or other reasons, the remaining engine still engages the central shaft and delivers torque to the main- and tail rotor. The inoperative engine must be disengaged, in order to remove any residual torque that has to be overcome by the remaining engine. The system is called *spragg clutch*. Other elements are indicated an the graph, but are not described further (rotor brake and auxiliary systems). The important point to take from this discussion is that the helicopter can continue to operate with a single engine, assuming that it can deliver the required torque until an emergency landing is performed.

An interesting case is the power train of the *Boeing-Vertol CH-47*, as displayed in Fig. 3.4. As the case previously discussed of the twin-engine conventional helicopter, there is a redundancy system to prevent a total collapse in case of one engine failure.

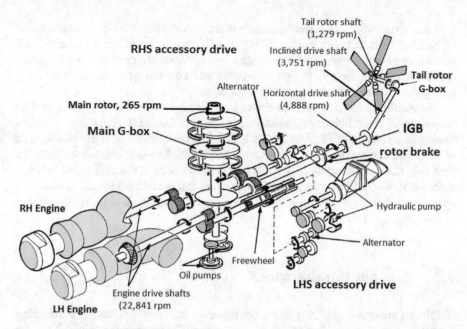

Fig. 3.3 Diagram of the transmission system of the Eurocopter/Airbus AS-332. Adapted from the Flight Crew Operating Manual

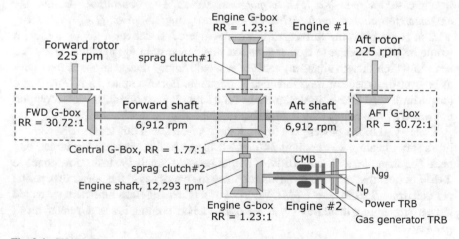

Fig. 3.4 CH47 helicopter drive train architecture; RR = reduction ratio

The architecture is based on the concept that two engines engage a central shaft, which then delivers torque to separate shafts. There is a spragg clutch per engine to allow disengagement, and a series of gear-boxes that reduce the engine shaft rpm to the much lower rotor 225 rpm. The total speed reduction is 1:54.64.

Among the twin rotor systems, notable is the power train of tilt-rotor aircraft, one embodiment of which is discussed by Wilson [2]. A tilt-rotor such as the Boeing V-22 is unlikely to be able to hover on a single engine. However, the central shaft system distributing torque through engine gear-boxes and a central gear-box dampens large torque imbalances and improves lateral stability.

Electrical tail rotors, and even multiple tail rotors have been proposed to take advantage of the efficiency of electrical machines. This new technology requires considerable changes in the transmission architecture, for example the inclusion of an electrical generator, power converters, electric cabling and control software. In principle, there are advantages in these new systems, such as the ability to disengage the tail rotor whilst idle on the ground (to save fuel and reduce noise), and the ability to operate at variable rpm, and thus optimise performance and reduce fuel consumption in level flight.

3.2.1 Gearbox Considerations

Gearboxes are critical and very heavy components, but there is no way of eliminating their presence, considering the very high speeds of the turboshaft engines. They are designed to deliver a maximum torque, and have a certified transmission limit, Q_{lim}. In any case, *the gear-box limit torque must exceed, by a reasonable amount, the maximum torque that can be delivered by the turboshaft engines, Q_{shaft}.*

If at some point the turboshaft are re-engineered to deliver a higher torque, it would be an advantage to have a gear-box that can sustain the upgrade. There is a limit to which this can happen, because it would imply that gear-boxes may have to be over-designed for their intended application. For example, the AW101 has a torque limit[1] of 112% of its design value for a a maximum continuous power output; this value increases to 118% for 2.5-min output rating.

Clearly, all of the power goes through the gear-box, and this creates problems in lubrication, heating and cooling. In fact, all losses in the gear-box are turned into heat. For example, with a 2000 kW of output power, a small 1% loss corresponds to 20 kW of heat that must go somewhere. Since the energy E is the power dissipation per unit time, $E = P \cdot t = 1.2$ MJ. With $\sim$ 10 kg of aero engines lubricant, we would have a temperature increase of $\sim$ 2–3 K/min. Hence, cooling is a requirement on all gear-boxes.

Some data exist in the technical literature on helicopter transmission weights [3], but this research is rather old (early 1980s), and is in need of an update. With the few data that are available in the open domain, it appears that the transmission weight grows linearly with the installed shaft power, Fig. 3.5.

---

[1] FAA Type Certificate Data Sheet H80EU, Revision 1, Feb. 2007.

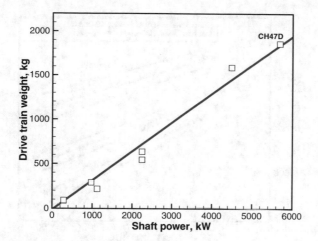

Fig. 3.5 Helicopter drive train weights (estimated)

3.3 Engine Power Ratings and Limitations

Turboshaft engines are certified by national and international bodies (FAA, CAA, EASA, US military, etc.) as other aero engines. They are characterised by different power ratings, that is necessary to specify. These power ratings will appear on the *Type Certificate*. Single-engine helicopters have turboshafts rated to power < 1000 kW.

The data included in this document are engine architecture, power ratings, engine speeds, engine limitations (including temperatures in sections of the engine), fuels and lubricants allowed, key dimensions, weight and reference to technical documentation. There are notable differences in engine type certificates between the FAA and the EASA, and between revisions, and it is always interesting to compare both in order to extract as much information as possible.

To begin with, engine manufacturers do not know what the installation issues are. The same engine can be installed on different helicopters with different intake configurations. There is some exchange of data between engine manufacturers and airframe manufactures, but there are restrictions. Thus, we have the *uninstalled* engine power. This is to be measured in reference conditions: sea level, standard day. Furthermore, some *Type Certificates* establish the specific conditions of the output ratings, which may include the fuel heat capacity, bleed extraction (if any), accessory loads, inlet pressure recovery, and exhaust discharge conditions. The key power ratings are:

- Maximum continuous power (MCP): this is the power that can be delivered indefinitely, e.g. for any length of flight, without measurable performance deterioration.
- Maximum take-off power (MTOP): this is a power output that can be delivered for 5 minutes, which is a general estimate for the duration of take-off and initial stages of climb-out.

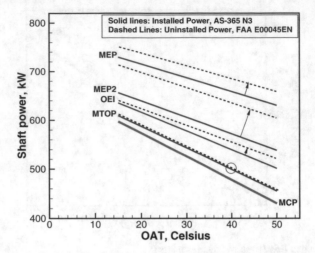

Fig. 3.6 Performance charts of the Turbomeca Arriel 2C2 turboshaft engine

- Emergency power rating (MEP): for a limited time (30 s, 2 min or otherwise), turboshaft engines can operate at increasing output to meet emergency situations. These events cause excessive heat, pressure and mechanical loads, and may need to be inspected before going back to service. This rating is sometimes called OEI rating.

In Fig. 3.6 we show the effects of engine power ratings, Outside Air Temperature (OAT), and installation effects, some of which will be discussed further. The documentation used for the preparation of this chart is indicated in the caption. For a fixed air temperature, note that the MCP is hardly affected by installation losses. These losses become materially important as we refer to shorter time frames and higher output ratings.

Another important effect is that of the air temperature. This engine (like any turboshaft) suffers from an increase in temperature (reduction in ambient air density), and on a hot day it can only deliver a fraction of its power on a standard day. Hence, hot day operations can be severely limited. Net out power depends on several factors, which include gas turbine discharge temperature (either TT4.5 or TT5), engine speed limits (N1%) and fuel flow limits (Wf6). This is illustrated in Fig. 3.7. The intermediate engine speed limit is quite narrow and now always clearly identifiable in engine test data. This behaviour is not shown in the data in Fig. 3.6, which are limited to higher OAT.

One engine can have military and civilian applications. For example, the Rolls-Royce/Allison M250 corresponds to the military engine T63. Furthermore, turboshaft can be developed into turboprops. For example, the General Electric T700 evolved into the CT7 turboprop (SAAB 340B), but there is a large series of engines within this family of turboshafts and turboprops.[2] Turboshaft engines evolve over a long time,

---

[2] EASA Type Certificate Data Sheet E.010.

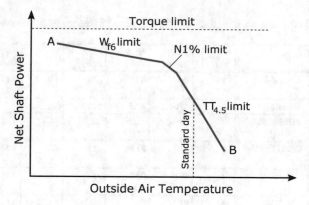

Fig. 3.7 Turboshaft engine power limits as function of outside air temperature. Turboshaft engine

and some of them have been in service for well over half a century (for example, the Lycoming T55), although the engine of today is a far relative of the original engine.

Limitations come in the form of limit speeds, temperatures and other operational constraints. For example, the N_{gg} speed is given as the nominal (design) 100% rpm, but it has upper limits of the order of 102–104%. Actual speeds can exceed 60,000 rpm on some small engines. There is generally a temperature limit in some section of the engine (inter-turbine temperature) or temperature at the exit of the last turbine stage, and the exhaust gas temperature (EGT). Operational constraints include minimum and maximum outside air temperature and flat rating data.

Data that are regularly missing include the overall pressure ratio, the design fuel flow rates, and the intake mass flow—all critical data for engine performance analysis. A few data are collected in Table 3.1. In the database shown, overall pressure ratios, Π, are below 17, and are notably lower than pressure ratios of modern high by-pass ratio engines. The main source of data for civilian helicopters is the Type Certificate, but for other engines one needs to carry out research across different databases.

From Table 3.1, we can infer some typical power/weight ratio (*specific power*): these range from just less than 2 kW/kg to over 10 kW/kg, growing rapidly with the engine power ratings. The trend is slightly biased toward newer engines. For example, the T53-L-701, one of the earlier versions of a family of turboshaft, appears slightly under-powered in comparisons with newer versions of the same family of engines, some of which have been converted into turboprops, Fig. 3.8.

3.3.1 Specific Fuel Consumption

There is no easy way to compare different turboshaft engines. One parameter that is often provided is the *specific fuel consumption*. This is defined as the ratio between the fuel flow and the net shaft power: SFC $= W_{f6}/P_{shaft}$. Contrary to common belief,

Table 3.1 Turboshaft engines data, static conditions, standard day, sea level; some data unavailable

Engine		W [kg]	TOP [kW]	MCP [kW]	MEP2 [kW]	Π	W_1 [kg/s]	Data
Turbomeca	Arriel 2C2	131	609	597	656	8.0	2.14	EASA E.001
Turbomeca	Arrius 2B2	112	479	432	554	8.4		
Turbomeca	RTM322-01/1	255	1518	1566	1669	14.2	4.52	EASA E.009
Rolls-Royce	C250 C20B/J	72	313	313	313	7.1	1.56	EASA E.052
Rolls-Royce	CTS-800-N	185	1014	955	1108	14.6	4.54	EASA E.232
General electric	T58-GE-8	139	912			8.3	5.61	
General electric	CT7-8	246	1879	1523	1879		> 20	EASA E.010
General electric	T64-GE-419	342	3570			14.9		
General electric	T700-GE-700	198	1210			17.0	4.5	
Rolls-Royce	T406(AE1107)	440	4586			16.7	16.0	Ref. [4]
Lycoming	T55-L-714A	377	3780	3620	3780	9.3		
Allison	T53-L-701	312	1082			7.4	4.85	
PZL	PZL-10W	143	662	574	846	7.0	4.50	EASA E.128

Weights sometimes quoted with/without reduction gear-box; data and/or type certificate sources in the right column. Where no data source is given, data are inferred from open source documents

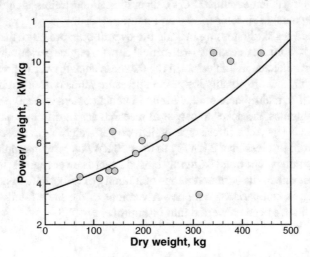

Fig. 3.8 Specific power data, P/W, for selected engines; power used is maximum continous power

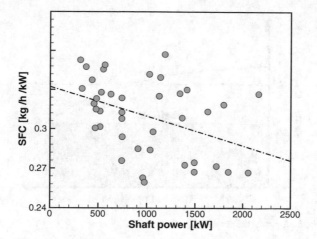

Fig. 3.9 Design SFC values for selected turboshaft engines (anonymised)

the SFC is *not a constant*. Its value depends on a large number of parameters and circumstances, and hence direct comparisons are impossible.

The physical units of the SFC are [mass/time]/[power]. In *International Units* (SI), SFC would be [kg/s/W], which is a rather small number. It is often converted to [kg/h/kW]. In the US, *Imperial Units* [lb/h/shp] are used, although these units are often neglected altogether in technical documentation and marketing brochures, and one needs to guess what the numbers mean.

A summary of SFC data available in the open domain is shown in Fig. 3.9. There is no evident correlation in performance, aside from a small improvements at high design power ratings. Is this sufficient to make an engine selection?

Aero engine performance depends on the jet fuel . Aviation fuel types are strictly regulated, and we refer to the relevant standard for their characteristics. For example, the commercial fuel Jet A1 has densities at 15 °Cin the range 0.775–0.840 kg/l, as reported by British Standards.[3] Military jet fuels include JP-4 and JP-5[4] At standard air temperature, the JP-4 has density of 0.751–0.802 kg/l, and JP-5 has a density of 0.788–0.845 kg/l.

Another important factor is the net caloric value, or the maximum chemical energy that can be converted to thermal energy. When fuel is burned, some energy is lost to heat in the water vapour, which is one of the products of combustion. In open systems such as turboshaft engines, this cannot be recovered. For this reason we use the "Lower Caloric Value" (LCV) of the fuel when establishing the required fuel flow rate, as opposed to the Higher Caloric Value (HCV) which assumes water is fully condensed. The LCV for Kerosene is 43 MJ/kg; the amount available for raising the temperature of the gas depends on the thermal efficiency of the engine,

---

[3] UK Ministry of Defence: Turbine Fuel, Kerosine Type, Jet A-1. DEF STAN 91 91, Issue 7, Amendment 3 (2015).

[4] Turbine Fuel JP-4 and JP-5: Standard MIL-DTL-5624U (1998).

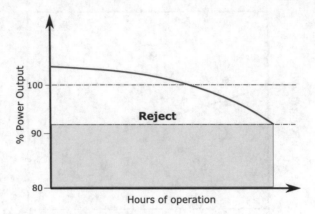

Fig. 3.10 Turboshaft power output deterioration

which is the ratio of useful work to the heat input during combustion. Typical values for turboshaft engines at design point are 0.25–0.35%, but depend on the operating altitude.

3.3.2 Engine Performance Degradation

When the engine is designed for a particular application, it is not unlikely that it is delivered with a slighter higher uninstalled shaft power. If we use normalised data, the engine manufacturer may be able to deliver an engine specification equivalent to 102% of the contracted power, interpreted for a *new engine, uninstalled, sea level, standard day*. Over time, the output power decreases for a number of factors, all related to deterioration effects in critical parts of the engine, due to thermal and mechanical fatigue, ingestion of particulate, mechanical erosion by lubricants or between metallic parts, etc. A power output curve may look like the one shown in Fig. 3.10. Depending on the customer, at 92–96% of the design output power would require an engine overhaul and maintenance, indicated as a "reject". In particulate-rich environments, the curve would feature a more rapid decrease in power output. When the rejection output is level, a compressor wash is triggered and some power out put is recovered, leading to a sawtooth shape in Fig. 3.10.

Engine rejection rates are often found to be much higher than predicted when operating in harsh environments such as desert conditions, littoral locations, and at sea. Mineral dust kicked up by rotor downwash or sea spray, contains silicates, carbonates, and sulphates, all of which attack substrates in different ways. Quartz is a very common mineral, and is particularly responsible for erosion of first stage compressor blades. However, this problem has been largely dealt with via the implementation of intake protection systems (see Sect. 3.5) and through application of ceramic-metallic matrix erosion-resistant coatings, albeit at a cost. But deposition of finer-grained material continues to pose several problems.

Fouling of compressor blades arises when sub-micron and micron-sized particles get trapped the thicker, slow-moving boundary layers towards the trailing edge of the pressure surface, and adhere under van der Waals force. Larger particles avoid this fate either due to stored elastic energy being greater than the attractive forces, or due to the centrifuging action of the rotating airflow leading to their complete avoidance of the blades. Fouling is a common problem for turbo-machinery, leading to a loss of aerodynamic efficiency and subsequent reduction in pressure ratio across a compressor stage for a given input of work. This can often be recovered through washing during regular maintenance visits.

In the hot section of the engine, where combustor walls and turbine vanes and blades are located, deposition causes more permanent damage. This happens in several ways. Firstly, deposited mineral dust can form an in-situ melt that permeates the interstices of thermal barrier coatings and attacks the underlying metal, leading to embrittlement and increased risk of crack formation. Secondly, molten material that has accumulated may solidify if the engine is run at a lower power setting (cooler core temperature). This can then lead to a spalling mechanism arising from the mismatch in thermal expansion coefficient between the two materials.

In each case, the deposit formation, and the associated performance deterioration, has been shown to exhibit an asymptotic exponential growth with mass delivered (see Ref. [5]). The time constant of this growth is dependent on the type of dust encountered, and the engine operating conditions (gas temperature, relative velocity), and reflects the trade-off between deposition and shedding. By fitting an asymptotic shape to the total pressure loss across a stage, we are able to obtain the time constant of deterioration, τ, and the asymptotic total pressure loss ω_∞, represented here as a coefficient:

$$\Delta\omega = \Delta\omega_\infty \left(1 - e^{-\frac{C_{1,c}}{N_b \tau}} \right) \tag{3.2}$$

where ω is the total pressure loss coefficient, $C_{1,c}$ is the dust concentration, and N_b is the number of blades or vanes per stage. This can be used to predict the rate of performance deterioration associated with particulate fouling and deposition.

3.4 Engine Performance Envelopes

Turboshaft engines are adversely affected by atmospheric conditions in a way that other gas turbine engines are not. The key effects are flight altitude (or pressure altitude) and air temperature.

A first-principle analysis is done by correcting fuel flow and shaft power for the effects of altitude, which materialise in density, pressure and temperature decreasing with the altitude (in standard atmosphere). This relationship is:

$$\frac{W_{f6}}{\delta\sqrt{\theta}} = A_e + B_e \frac{P}{\delta\sqrt{\theta}} \tag{3.3}$$

where the coefficients A_e and B_e depend on the engine, P is the total shaft power. If there are n engines, then Eq. 3.3 will have the term P/n. At zero net shaft power, the fuel flow is $W_{f6} = A_e$. If both coefficients are constant, then there is a linear relationship at all altitudes between fuel flow and shaft power.

3.4.1 Engine Simulation Models

Engine simulation offers a wider possibility for exploring the engine envelope than any flight testing. It is faster, far less expensive and risky. The main drawback is the lack of reference data and validation procedures. With these caveats in mind, in recent years there has been a considerable amount of development in the area of aero-thermodynamic simulation of gas turbine engines. A number of sophisticated computer codes are available, such as the NASA Propulsion System [6, 7] (NPSS), the Gas Turbine Program GSP [8, 9], TurboMatch [10] and a number of other commercially available computer codes (for example, *GasTurb*).

One common feature of these computer codes is that they model the one-dimensional (gradients only along the gas path) gas flow across all the engine sections, both steady and unsteady, and are able to provide a sufficiently detailed map of the engine performance. A comparison between computer models has seemingly not been made available in current publications, but it is likely that some differences exist, not necessarily due to the core models, but also due to the use and integration of the input data. A model of the engine is shown in Fig. 3.11. The key systems of the engines are numbered from 1 (inlet) to 9 (exhaust).

Key temperatures are given at the exit of each system, $T_1 \ldots, T_5$, with TIT the inter-turbine temperature. The TIT is sometimes called $T_{4.5}$ (the exit of the high-pressure turbine) or $T_{4.6}$ (entry to low-pressure turbine).

The TIT or the total temperature at the exit of the last turbine stage T_5 are given as operational limits in the engine's *Type Certificate*. The temperature $T_4 > 1800\,\text{K}$ and the exhaust gas temperature (EGT) can be in well in excess of $500\,\text{K}$. The exhaust turbine temperature has limits between 550 and $750\,\text{K}$, from idle to take-off power, respectively.

One example of comparisons between different simulations is shown in Fig. 3.12. The lapse rate of the engine power is closer to $\sigma^{0.8}$ for the case of the GSP simulation and σ for the NPSS simulation (σ is the relative air density on a standard day). However, there are glaring differences between the two simulated data sets. The relationship $P(z) = P_{sl}\sigma^{0.8}$ is often used in first-order altitude analysis of rotorcraft.

Figure 3.13 shows a comparison between two different models of the T700 turboshaft engine and some reference data [11]. In this case we show the net torque at the engine shaft as a function of the fuel flow. The Q-W_{f6} only shows a constant bias (except the first point) on the reference data.

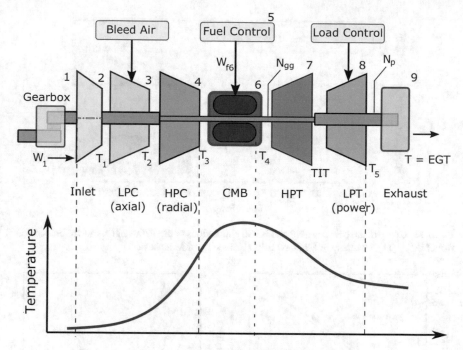

Fig. 3.11 One dimensional aero-thermodynamic model of a turboshaft engine

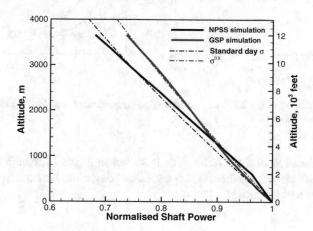

Fig. 3.12 Altitude-dependent shaft power of a model T700 turboshaft engine, simulated with GSP and NPSS

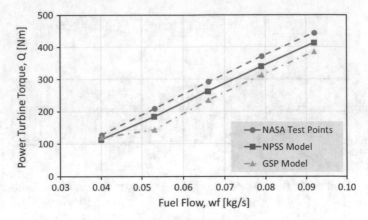

Fig. 3.13 Comparison between GSP and NPSS simulations of the T700 engines and reference data from NASA [11]. Courtesy of Matthew Ellis (University of Manchester)

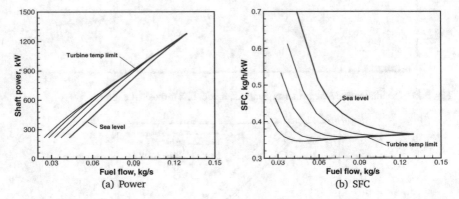

Fig. 3.14 Simulation of the T700 turboshaft engine on a standard day. Power and SFC lines at 2000 m/6560 feet intervals

A full simulation of the T700 engine is shown in Fig. 3.14, which displays the unistalled shaft power and the SFC on a standard day. Note how the SFC increases rapidly at low fuel flow rates.

3.4.2 Engine Emissions

Turboshaft engines do not fall within the ICAO regulations of emissions, and hence there is no obligation for engine manufacturers to report emissions characteristics as it is done for large turbofan engines. However, the exhaust emissions are equally important. Lacking rational reference data, we can only refer back to the sparse studies available in the technical literature, one of which, Ref. [12] clearly highlights

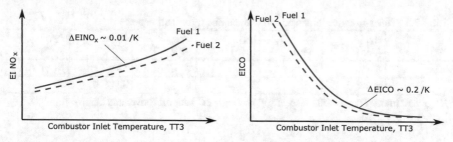

Fig. 3.15 Notional behaviour of some exhaust emission indices for a turboshaft engine

the difficulties of assessing the emissions for this type of vehicles, but it also proposes ways of providing estimates.

Emissions data that are of interest are CO (carbon-monoxide), NO_x (nitrogen oxides), UHC (uncombusted hydrocarbons), SO_x (sulphur oxides), non volatile particulate matter (mostly black carbon) and volatile organic compounds (VOC). Emissions of CO_2 are referred back to the fuel burn, since there is a constant multiplier between the fuel burn and the carbon emissions. This multiplier depends on the fuel and fuel-air ratio, but it is generally assumed to be 3.156 kg/kg of fuel.

Emissions other than CO_2 depend on fuel burn in non linear relationships. In fact, other factors intervene, in particular the entry conditions in the combustor, the maximum temperature and pressure in the combustor and the geometry of the combustor itself. At least one study of the emissions of the T700 turboshaft engine is available [13] and one on the T63 turboshaft [14]. A general trend of emission indices for NO_x and CO is displayed in Fig. 3.15. The data shown are emission indices, normally given as [g/kg] of fuel burned. Difference between aviation fuels have also been reported.

NO_x emissions increase with combustor temperature. Increasing combustor temperature in modern engines, due to higher compression ratios has caused this side effect. The opposite has taken place on the carbon-monoxide emissions, which is the product of incomplete combustion.

Emission assessments for helicopters is a far less formalised process than the ICAO-regulated turbofan engines. In part, this is due to the complexity of airborne operations carried out by this type of vehicles; in part we have a notable absence in emissions standards for this type of engines.

In order to begin making an assessment, we need to have average fuel flow rates and *time-in-mode* t_i estimates. A time-in-mode is the amount of time the engine runs at a given rpm, for example during idle, take-off, climb-out and cruise. The fuel burn for each mode i is $W_{fi} = W_{f6_i} t_i$:

$$W_f = \sum_i W_{f6_i} t_i \qquad (3.4)$$

Table 3.2 Times-in-mode for single and twin-engine turboshaft engines

	t_{ID} [min]	t_{AP} [min]	t_{TO} [min]	P_{ID} %	P_{AP} %	P_{TO} %
Single	5	3.5	3	7	38	78
Twin	5	5.5	3	6	32	66

ID ground idle; *AP* approach (descent and landing); *TO* take-off (hover and climb)

Values suggested for times-in-mode are indicated in Table 3.2, adapted from Ref. [12]. For each of the operating modes (ground idle, approach, take-off), we need an estimate of the fuel flow (which depends on the shaft power), and finally an estimate of emission indices for each chemical emission species.

We assume that the fuel flow to power relationship can be established by numerical simulation with one of the gas turbine methods described in Sect. 3.4.1, otherwise some empirical estimates are available. Hence, there is a suggestion of empirical relationships derived from testing, as given below:

$$
\begin{aligned}
&\text{EI}_{NO_x} \simeq 0.2113\, P_s^{0.5677} && [\text{g/kg}] \\
&\text{EI}_{UHC} \simeq 3819\, P_s^{-1.0801} && [\text{g/kg}] \\
&\text{EI}_{CO} \simeq 5660\, P_s^{-1.1100} && [\text{g/kg}] \\
&\text{EI}_{nvPM} \simeq 0.1056 + 2.3664 \times 10^{-4}\, P_s - 4.8 \times 10^{-8}\, P_s^2 && [\text{g/kg}] \\
&\text{PM} \simeq \text{EI}_{nvPM} / \tfrac{\pi e^{2.88}}{6} \left(\overline{d}_p^3 \right) && [-]
\end{aligned}
\tag{3.5}
$$

where the power P_s must be expressed in shaft-horsepower SHP (with a conversion factor 1 kW = 0.7457 SHP) and $\overline{d}_p$ is the mean particle diameter [nm]. At this point, it is possible to make a very approximate estimate of the LTO emissions. For example, for the NO_x we have

$$
\text{LTO NO}_x = 60\, n E I_{NO_x} \left(t_{ID} W_{f6_{ID}} + t_{TO} W_{f6_{TO}} + t_{AP} W_{f6_{AP}} \right) \quad [\text{g}]
\tag{3.6}
$$

since the fuel flow is in [kg/s] and the times-in-mode are given in minutes. For each operational mode, calculate the required power P; from the power, calculate the fuel flow W_{f6}; the fuel burn is calculated from the time-in-mode, and the corresponding emission of a chemical species is calculated from multiplying the emission index, Eq. 3.5, by the fuel burn in that mode. This procedure is further described in Sect. 3.6 below.

The procedure is clearly approximate because the emission indices only appear to depend on the shaft power and not on the specific engine, jet fuel, deterioration effects, as indicated in Fig. 3.15. However, with these approximations, we have the general behaviour illustrated in Fig. 3.16, which shows the case of a notional 1200 kW turboshaft engine. When the helicopter is operating in idle mode on the ground ($P \simeq$ 90 kW), emissions of carbon monoxide and uncombusted hydrocarbons are more than four times higher than the corresponding emissions at other operational modes.

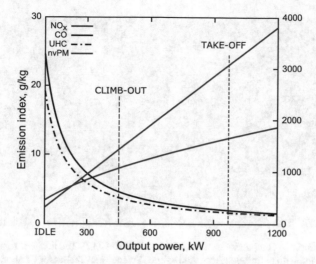

Fig. 3.16 Behaviour of some exhaust emission indices as a function of shaft power

This result is in agreement with complaints of poor air quality where helicopters operate near the ground. Some emission analysis methods have been published in the technical literature, Ref. [15] and some include the effects of the jet fuel composition [16].

A more detailed process for estimating emissions requires a mission analysis (Chap. 4). Mission optimisation can be carried out to minimise fuel burn and emissions. An alternative approach is to look at databases of flight trajectories for rotorcraft that use the standard ADS-B transponders, and simulate or 're-analyse' the trajectory using an accurate engine model.

3.5 Engine Intake Protection Systems

An intake protection system prevents particulate, dust and foreign object debris from being ingested into the engine, which can potentially have catastrophic consequences on the engine. This can be instantaneous damage or long deterioration or erosion of critical parts in the engine. An example of compressor blades erosion is shown in Fig. 3.17.

The amount of dust raised correlates directly to the average disk loading, and to a lesser degree the shape of the wake (number of trailing vortices—see Ref. [18]). For example, it was reported in Ref. [19] that dust concentrations at the rotor tips, at a height of 1.4 m, varied from 310 mg/m^3 for a Bell UH-1 to 2110 mg/m^3 for a Sikorsky MH-53. The corresponding disk loadings were 24 and 48 kg/m^2. Thus, a doubling of the disk loading caused a massively higher dust cloud concentration. Under these conditions, engine damage is inevitable without the use of intake protection. It must

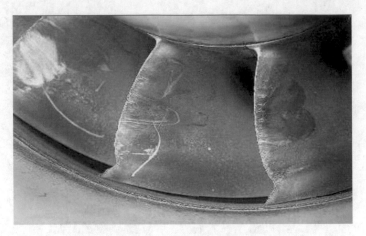

Fig. 3.17 Example of compressor erosion, adapted from Ref. [17]. The image is reproduced under Crown Copyright, with permission from the Defence Science and Technology Laboratory (DSTL). DSTL is an executive agency of the UK Ministry of Defence (MoD)

be noted that concentration increases rapidly toward the ground (3200 mg/m^3 for the MH-53 at a ground height of 0.5 m). This is not the dust that will be ingested: concentration, particle size distribution depend on operational conditions and chemical composition depends on the ground conditions. In order to prevent such dust ingestion, helicopters operating on these harsh environments are equipped with intake particle separators. There are three types of systems, of which a full technology review is published in Ref. [20]:

1. Inertial Particle Separators (IPS): these separators can be an integral part of the intake or installed as add-ons in front of the engine. These systems are rather common, at least for military engines.
2. Vortex Tube Separators (VTS): these are packs of helix-shaped channels in front of the engine, which separate the air flow on the principle of inertia (as above), although particle collision dynamics plays an important part.
3. Inlet Barrier Filters (IBF): as the case of the vortex tubes, these are add-on separators that are installed at the inlet, and consist of fiber-based filtration principles.

For all separators, we have two very important performance parameters: the separation efficiency and the pressure loss at the inlet. Inevitably, there is a cross-correlation between these parameters and an engineering compromise is required. Separation efficiencies η_s are as follows: inertial particle separator $\eta_s \sim 80\%$; vortex tube separator: $\eta_s \sim 95\%$; inlet barrier filter $\eta_s \sim 99\%$. These data are based on a coarse test dust with particle diameters 2–200 μm, and a mean diameter $d_p = 38$ μm.

The pressure loss at the inlet is either depending on the operating point (IPS, VTS), or growing with time (barrier filters). In the latter case, since there is a continuous accumulation of particulate within the filter layers, the free passage of the air flow into the inlet is obstructed, and this is the cause of increasing pressure losses. A partial recovery of the pressure loss is achieved by cleaning and washing of the filter at specified intervals. A loss of inlet pressure requires the compressor to work harder in order to deliver the required pressure ratio in the combustor.

Pressure losses across a clogged filter can be as much as a few kPa if the rotorcraft engines rapidly ingest fine talc-like dust. In such circumstances, typically during a *brownout* or *dust landing* it may become necessary to operate a bypass system, in order to avoid stalling the compressor. The pressure loss, δP_{IPS}, follows a quadratic with engine mass flow rate, or dynamic pressure, q:

$$\delta P_{IPS} = c_1 q + c_2 q^2 \tag{3.7}$$

where c_1 and c_2 are empirical factors that depend on the geometry of the separator. In the case of inlet barrier filters, these are non-constant and depend on the degree of clogging.

In the diagram of Fig. 3.2, there in an indication of an IPS presence. In this case, the IPS is bolted in front of the engine, to the left in the drawing. With a high-speed IPS particles and gas are forced around a hump. The particulate would tend to follow the path of lower duct curvature and is thus separated by inertia. Larger particles, that enter the intake in a more ballistic fashion, may reach the scavenge by a favourable sequence of interactions with the duct walls, although the same mechanism has been known to lead to the inadvertent ingestion of sand and gravel.

These systems were first developed in the 1970s by GE for the T700 engine; more recently, they have been applied to the RTM322, Fig. 3.2, and other engines. Large flow speeds into engine (70–90 m/s) are possible, but to centrifuge particles particles with diameters $d_p < 25\,\mu m$ requires that the flow accelerates rapidly, both linearly and angularly, which may result in large pressure losses, of the order ~ 1 kPa. Separation efficiency is good for particle diameters $d_p > 25\,\mu m$.

The particulate is scavenged radially by a pump, leaving a cleaner core gas into the engine. To cater for the increased mass flow rate, the intake area is enlarged. The use of the pump requires additional power, which is taken from the engine. The power loss can be as much as 10%. Clearly, this causes an overall loss in propulsive efficiency, since a small part of the work has to be done to provide clean air. For this reason, particle separators must only be used when operating in harsh environments; they are not a standard piece of equipment. *A life-cycle analysis is required for an overall cost-benefit assessment.* Additional costs that would be considered in such an analysis include: system weight, maintenance burden, drag, reliability, integration, physical envelope.

3.6 Airframe-Engine Integration

There are considerable issues in the integration of a power plant onto an helicopter aircraft. We only consider the operational cases, when computer simulation is required. This is illustrated in the flowchart of Fig. 3.18.

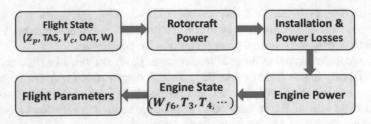

Fig. 3.18 Flowchart for calculating the engine-airframe integration

To begin with, we need to establish the flight conditions. These include at least the gross weight, flight speed, climb rate, flight altitude, atmospheric temperature, and possibly other parameters. This flight state is used to calculate the gross power required to fly at that condition, in trimmed state. The actual engine power depends on a number of integration factors, including transmission losses, intake and exhaust losses, presence of particle separators, and ageing effects (see also Fig. 3.10). Since there is often a need of compressor bleed air, that contributes to a relative reduction in air flow, or the reduction in net shaft power—this is an installation effect, which can be anything up to 20% of total power. Other effects include exhaust back pressure, due to friction or otherwise (of the order of 1%) and in case of military vehicles with infra-red suppression systems there is also a back pressure loss due to this additional system (in greater measure, presumably up to 15% of power).

At this point, there is a power/torque requirement, and the engine gas dynamics problem is solved in inverse mode. This means that we need to trim the fuel flow W_{f6} to deliver the required shaft power. Once a solution is achieved, that corresponds to an engine state that has several aero–thermodynamic parameters of interest: temperatures and pressures at critical sections of the engine, mass flow rates, etc. Finally, flight parameters are calculated, such as the specific air range and the specific endurance.

3.7   Case Study: Optimal Operation of a Multi-engine Helicopter

An interesting case for optimal cruise performance with one engine inoperative has been proposed in Ref. [21] for the three-engined AW101 *Merlin* helicopter. The following case study is based on the approach that it is possible to gain fuel efficiency at some flight conditions with one engine inoperative.

We discussed the features of twin-engine rotorcraft, but very little was reported on three-engine configurations. Now consider the latter case. We have a three-engined rotorcraft powered by RTM322 turboshafts. Flying at the loiter speed (approximate), is there any gain in disengaging one engine for the purpose of improving fuel flow characteristics?—What is the effect on the SFC?

(a) Mil-14

(b) Mil-17

Fig. 3.19 Examples of three-engined helicopters: Mil-14 in search-and-rescue configuration and Mil-17 military aircraft. Photos courtesy of Filip Modrzejewski (2019)

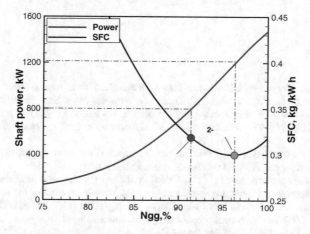

Fig. 3.20 Twin- versus three-engine turboshaft operation

Photographs of three-engined rotorcraft are shown in Fig. 3.19. In the case of the Mil-17, note the inertial particle separators on the two lower engines. The central higher engine is more protected from the ingestion of dust and particulate and is thus not equipped with this system.

For the purpose of this discussion, assume that at this flight condition (steady level flight) we require a total of 2400 kW, which delivered in equal measure by the three engines (800 kW/engine). The performance curve is shown in Fig. 3.20, which displays both power and SFC as a function of the gas turbine speed, N_{gg}. The data have been generated by one of the engine simulation programs mentioned earlier (GSP). We demonstrate that in this case there is some efficiency gain in operating the rotorcraft as a twin-engine—with some caveats.

To begin with, the three-engine operation would require $N_{gg} \simeq 91.5\%$ to meet the 800 kW output power demand, with an SFC $\simeq 0.32$ kg/h/kW. The twin-engine operation would require a power output 1200 kW/engine. This is achieved with $N_{gg} \simeq$ 93%, although it has the benefit of *decreased* SFC to ~ 0.30 kg/h/kW. This means a reduction in SFC by over 6%, which is considerable. A similar analysis could be carried out for twin-engine rotorcraft operating as a single engine [22], since there are fuel consumption benefits at higher engine loads.

Operating at higher rpm may cause an increase in direct operating costs, particularly on the maintenance side of the costs, and this is not evidenced in the graph, because the graph only shows the effects of fuel burn. Furthermore, the safety implications are not discussed, and these need to be carefully addressed, if there is a sudden need of a power supply. It is well known that the restart time of a gas turbine engine can be several seconds, and in emergency situation there can be loss of speed and altitude; this limits the safe envelope of the intended OEI flight. We conclude that a *complete analysis of engine performance needs to take into account factors that are beyond fuel burn.*

References

1. O'Brien DM, Calvert ME, Butler SL (2008) An examination of engine effects on helicopter aeromechanics. In: AHS specialist conference on aeromechanics, San Francisco, Jan 2008
2. Wilson HK (1991) Drive system for tiltrotor aircraft. US Patent 5,054,716, Oct 1991
3. Weden CJ, Coy JJ (1984) Summary of drive-train component technology in helicopters. Technical report TM 83726, NASA, Oct 1984
4. Arvin JR, Bowman ME (1990) T406 engine development program. In: ASME gas turbine & aeroengine congress, ASME 90-GT-245, Brussels, June 1990
5. Döring F, Staudacher S, Koch C (2017, June) Predicting the temporal progression of aircraft engine compressor performance deterioration due to particle deposition. In: Volume 2D: Turbomachinery. ASME, p V02DT48A007. https://doi.org/10.1115/GT2017-63544
6. Lytle JK (2000) The numerical propulsion system simulation: an overview. Technical report NASA TM-2000-209915, NASA, June 2000
7. Jones SM (2007) An introduction to thermodynamic performance of aircraft gas turbine engine cycles using the numerical propulsion system simulation code. Technical report NASA TM-2007-214690, NASA, Mar 2007
8. Visser WPJ, Broomhead MJ (2000) GSP: a generic object-oriented gas turbine simulation environment. In: ASME gas turbine conference, ASME 2000-GT-0002, Munich, Germany. Program available from www.gspteam.com
9. Visser WJP, Kogenhop O, Oostveen M (2004) A generic approach for gas turbine adaptive modeling. J Eng Gas Turbines Power 128(1):13–19. https://doi.org/10.1115/1.1995770
10. Pachidis V, Pilidis P, Marinai L, Templalexis I (2007) Towards a full two dimensional gas turbine performance simulator. Aeronaut J 111(1121):433–442. https://doi.org/10.1017/S0001924000004693
11. Duyar A, Zu G, Litt JS (1995) A simplified model of the T700 turboshaft engine. J Am Helicopter Soc 40(4):62–70
12. Rindlisbacher T, Classey L (2015) Guidance on the determination of helicopter emissions. Technical report, FOCA, Bern, Switzerland, Dec 2015
13. Cohen JD (1984) Analytical fuel property effects—small combustors. In: Grobman J (ed) Assessment of alternative aircraft fuels, NASA CP 2307, pp 89–98
14. Kinsey JS, Corporan E, Pavlovic J, DeWitt M, Klingshirn C, Logan R (2019) Comparison of measurement methods for the characterization of the black carbon emissions from a T63

turboshaft engine burning conventional and Fischer-Tropsch fuels. J Air Waste Manag Assoc 69(5):576–591. https://doi.org/10.1080/10962247.2018.1556188

15. Ortiz-Carretero J, Castillo Pardo A, Goulos I, Pachidis V (2018) Impact of adverse environmental conditions on rotorcraft operational performance and pollutant emissions. ASME J Eng Gas Turbines Power 140(2). https://doi.org/10.1115/1.4037751

16. Li-Jones X, Penko P, Williams S, Moses C (2007) Gaseous and particle emissions in the exhaust of a T700 helicopter engine. In: Volume 2: Turbo expo 2007 of turbo expo: power for land, sea, and air, 5 2007. https://doi.org/10.1115/GT2007-27522

17. Martin N (2019) Gas turbine engine environmental particulate foreign object damage. Technical report STO-TR-AVT-250, NATO. https://doi.org/10.14339/STO-TR-AVT-250

18. Milluzzo J, Gordon Leishman J (2010, July) Assessment of rotorcraft brownout severity in terms of rotor design parameters. J Am Helicopter Soc 55(3):32009–320099

19. Cowherd C (2007) Sandblaster 2 support of see-through technologies for particulate brownout—task 5. US Army Aviation & Missile Command, Technical report, Oct 2007

20. Bojdo N, Filippone A (2010) Turboshaft engine air particle separation. Prog Aerosp Sci 46:224–245

21. Massey C, Violante T, Highams L (2018) One engine inoperative cruise for mission performance optimisation. In: Proceedings of the 74th annual vertical flight society forum, Phoenix, AZ, May 2018

22. Kerler M, Erhard W (2014) Evaluation of helicopter flight missions with intended single engine operation. In: European rotorcraft forum, Sept 2014

Prof. Antonio Filippone is at the School of Engineering, University of Manchester, where he specialises in aircraft and rotorcraft performance, environmental emissions aircraft noise, and engine performance.

Dr. Nicholas Bojdo is at the School of Engineering, University of Manchester, where he specialises in aero engine propulsion, intake design, engine air particle separator systems, dust and volcanic ash ingestion, engine deterioration.

Chapter 4
Rotorcraft Flight Performance

Antonio Filippone

Abstract This chapter introduces rotorcraft steady-state flight performance and stability, and we explain key concepts of the conventional helicopter as well as other rotorcraft types (tandem helicopter and compounds). Numerical models of performance estimations are provided for level flight, climb-out, and descent. Stability issues presented include longitudinal/lateral trim and speed stability. Different take-off procedures are illustrated, alongside the certification requirements (Category A and B rotorcraft). There is further discussion of ground effects, such as lift augmentation and ground resonance. We provide examples of methods used to estimate the direct operating costs, which are one of the major limiters to the use of rotorcraft. We complete the performance analysis with fuel planning methods, payload-range assessment, and speed augmentation concepts (compound helicopters).

Nomenclature

AEO	all engines operating
AI	autorotation index
OEI	one engine inoperative
SAR	specific air range
SEP	specific excess power
SFC	specific fuel consumption
TAS	true air speed
A, B	empirical coefficients in ground effect equation
A	rotor disk area
c	wing or blade chord
C_{DV}	vertical-flow drag coefficient
C_{D_g}	drag coefficient in ground effect
$C_{L\alpha}$	lift-curve slope

Supplementary Information The online version contains supplementary material available at https://doi.org/10.1007/978-3-031-12437-2_4.

A. Filippone (✉)
School of Engineering, The University of Manchester, Manchester, UK
e-mail: a.filippone@manchester.ac.uk

C_M	pitching moment coefficient
C_T	thrust coefficient
$C_{T\sigma}$	effective blade loading coefficient, C_T/σ
$C_{y\beta}$	side force derivative
D	drag force
D_L	disk loading, T/A
I_b	blade's polar moment of inertia
$k_1 \cdots k_4$	stability derivatives, Eq. 4.25
k	blockage factor
k_g	ground effect factor, Eq. 4.16
h	rotor height above ground
L	lift force
M	true Mach number
n	number of operating engines
P	power
q	dynamic pressure
Q	torque
R	rotor radius
$\mathcal{R}$	rolling resistance
t	flight time , s
T	rotor thrust
T_p	propeller thrust
V	flight speed
W	gross- or take-off weight
W_1	corrected weight, due to rotor downwash
W_{f6}	fuel flow per engine , kg/s
x_{cg}	distance between main rotor and centre of gravity
x_{ht}	distance between H-stabiliser and centre of gravity
x_{tr}	distance between main- and tail rotor shafts
X	flight range, n-miles
Y_{vt}	side force of the vertical stabiliser
z	flight altitude (m or feet)
$\bar{z}$	normalised ground clearance, z/D

Greek Letters

α	inflow angle
α_r	rotor disk tilt angle
β	sideslip angle
δ	differential
γ	ratio between specific heats
λ_i	inflow velocity ratio
φ	tail rotor cant angle

μ rotor advance ratio
μ^* normalised rotor advance ratio, $\mu\sqrt{2/C_T}$
μ_r ground rolling coefficient
σ rotor solidity
θ collective pitch angle
ρ air density
Ω rotor angular speed

Subscripts/superscripts

$\overline{(.)}$ mean value
$(.)_a$ airframe
$(.)_{cr}$ critical value
$(.)_i$ induced
$(.)_j$ iteration count
$(.)_{ige}$ in ground effect
$(.)_{oge}$ out of ground effect
$(.)_{gbox}$ gearbox
$(.)_{mr}$ main rotor
$(.)_p$ payload
$(.)_{ht}$ horizontal stabiliser
$(.)_{ref}$ reference value
$(.)_{tr}$ tail rotor
$(.)_{vt}$ vertical stabiliser
$(.)_w$ wing
$(.)^*$ corrected value

4.1 Introduction

The value of the helicopter is determined by what an operator can do with it. Typical questions are: How fast can it travel?—How far can it go?—How long can it stay airborne?—How much payload can it carry?

The issue of rotorcraft speed has been around since the first generation of rotorcraft, and it has been emphasised that speed is not the best virtue of the helicopter.

The ability to stay airborne motionless (hover) is indeed a virtue, since it allows the vehicle to perform operations such as search and rescue, precise delivery or collection of payload, raising and lowering sling loads, and a variety of other missions. Where the helicopter falls short, is overall endurance, e.g. the amount of time it can remain airborne, since this is seldom longer than 2 h. In the following sections we will review the performance characteristics at the most important flight segments, which are identified as follows: hover, vertical flight, climb-out, loiter, cruise, descent, weight-drop, weight-load, accelerate, decelerate, turn, etc. Figure 4.1 shows an example of

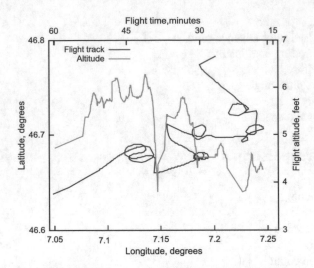

Fig. 4.1 Flight track and altitude of a search or recoinnassaince helicopter operation

flight operation for a certain helicopter on a typical day. The data have been extracted from an ADS-B database, and show show in this instance the helicopter performs several loops around a search area and changes altitude almost continuously.

When we look at the rotor performance, the tip speed must maintain a range between a minimum of stored kinetic energy (for autorotative performance require-ments) and noise limitations—at all flight speeds, as explained in Fig. 4.2. As the flight speed increases, the rotor suffers one of two problems: either compressibility effects, due to high transonic Mach numbers of the advancing blade, or dynamic stall limitations, due to stall of the retreating blade, severe stall on the advancing blade, both compounded by large pitch oscillations. At point B in the graph, both aerodynamic compressibility and dynamic stall contribute to constraining the flight speed. Note that all these limitations are reached even before we involve issues of engine power. In fact, in most cases engine power is not a limiter to helicopter speed.

There must be ways to make rapid assessments of the helicopter capability by using first-order analysis. For example, the disk loading W/A is an indicator of the gross rotor loading. At maximum weight, the thrust delivered must be well in excess of the weight. High disk loading is related to the strength of the rotor downwash, which dictates what kind of ground operations are feasible from a safety point of view. Light utility helicopters have the lowest disk loading, and tilt-rotors (convertiplanes) have the largest loading.

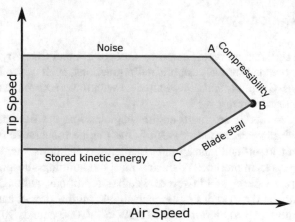

Fig. 4.2 Tip speed versus helicopter speed

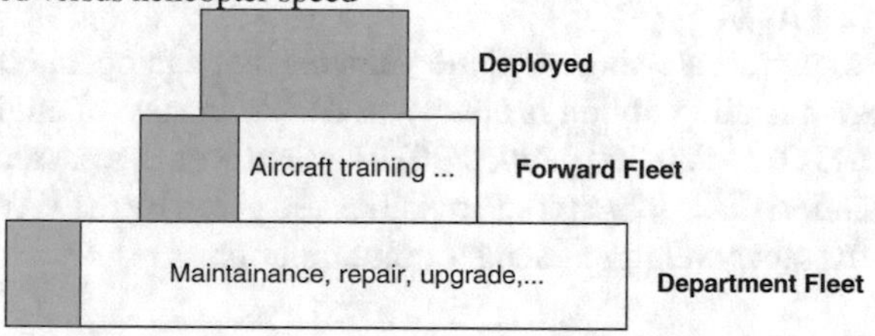

Fig. 4.3 Example of military rotorcraft reliability, availability and maintenance map

4.2   Reliability, Availability and Maintenance

No single air vehicle can operate continuously over a long period of time (measured in hours or days or weeks), due to the complexity of the logistics that is required to supply routine maintenance, spare parts, etc. This can be particularly critical for the military. Only a small fraction of the helicopters in the fleet are operational at any given time. This is indicated as "deployed" in the graph of Fig. 4.3. The remaining large portion of the fleet is either a "forward fleet", mostly used for training purposes, or unavailable, because under maintenance, upgrade, repair, testing, in storage, etc. The case shown is typical of many military forces, with only ∼1/5 rotorcraft in the fleet that can be deployed at any given time. This case indicates a very complex and costly supply chain. Thus, efforts are required to make the pyramid steeper (or even revert it), by increasing the number of deployable vehicles, reducing the the vehicles used for training purposes (through flight simulators), reducing the need for maintenance and repair However, the problem of affordability remains, since very often the installation of every kit of equipment (avionics, systems, weapons) on every vehicle is not affordable, and perhaps not even needed.

4.3 Certification Types

There are two certification types, noted as Category A and Category B. Transport rotorcraft are governed by well established regulations, such as the Federal Aviation Regulations, Chap. 14, Part 29, which deal with the *airworthiness standards of transport category rotorcraft*.

Category A rotorcraft are multi-engine vehicles that are designed operate in demanding environments, even in cases of One Engine Inoperative (OEI). Specifically, they can take-off and land from/to challenging heliports, including offshore platforms, ship decks in presence of shear winds, tall buildings, hospitals. They can operate in narrow valleys, and in severe weather conditions, although the level of severity depends on the avionics systems available on the aircraft and the training levels of the pilots. All rotorcraft with MTOW > 20,000 pounds/9072 kg, and 10 or more passengers, must have a *Type A* certification, regardless of the number of engines (FAR §29.1(c)).

Category B rotorcraft are single-engine vehicles that can operate from relatively safe areas, where a landing option is always available in case of engine failure. All rotorcraft with MTOW < 20,000 pounds/9072 kg, and 9 or less passengers, can have a *Type B* certification (FAR §29.1(f)). Provisions are given by the FAR §29 for cases not listed here for other weight/passenger combinations.

4.4 Point Performance Parameters

A point-performance index is a parameter that depends on the instantaneous operational conditions of the rotorcraft: flight altitude, air speed, climb rate, gross weight and atmospheric conditions. As one of these operational conditions change, so doe the point performance.

The specific air range (SAR) is the distance that can be flown by burning one unit of fuel. This is either expressed in unit volume or unit mass (or weight, to add confusion). In this instance it is defined as the true airspeed divided by the fuel flow

$$\text{SAR} = \frac{V}{W_{f6}} = \frac{1}{\text{SFC}} \frac{V}{P_{shaft}} \tag{4.1}$$

Note that in this case we use the fuel flow in kg-weight, as this is common engineering practice. A preferred unit for the SAR is [m/kg], or [n-mile/lb] in imperial units. Often a hybrid unit is used, such as [n-mile/kg]. This parameter is only useful when the helicopter flies in level flight. In other flight conditions, its use is inappropriate, and in hover it is meaningless, because $\text{SAR} = 0$.

To determine hover performance we use is the specific endurance, defined as the time required to burn one unit of fuel. Again, we use the unit of mass, so we have

$$E = \frac{1}{W_{f6}}$$ (4.2)

Specific endurance is expressed in seconds, because fuel flows are of the order of a fraction of a [kg/s].

The specific excess power (SEP) is

$$SEP = \frac{P_e - P_{total}}{W}$$ (4.3)

e.g. the ratio between the net excess power and the gross weight. The excess power is the difference between what the engines can deliver at the required operational point and the power required to fly at at that point. Note that the physical dimensions of this quantity are [m/s], and thus it is a velocity, not a power. This parameter will be used in the discussion of climb performance, Sect. 4.8.

The specific fuel consumption (SFC), discussed in Chap. 3, is also a point performance, because its value changes with all the operational parameters listed.

4.5 Take-off and Landing Procedures

We discuss a number of ground operations intended to comply with the safety regulations of Category A rotorcraft. The actual take-off depends on whether there is a landing alternative in case of emergency requiring the flight to be aborted. For cases where such an alternative does not exist (take-off from helipads on off-shore bases, building tops, etc.), a short vertical climb is followed by a backward climb-out to a target altitude leaving the helipad slightly in front of the rotorcraft. At this target altitude, the rotorcraft pitches down, loses some altitude whilst accelerating and then climbs-out to its target cruise altitude and speed. If an engine failure occurs, the rotorcraft must demonstrate the ability to clear a minimum height of 35 ft above ground/sea in the most unfavourable conditions. This operation is displayed in Fig. 4.4.

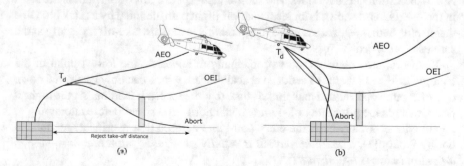

Fig. 4.4 Take-off procedures of Category A rotorcraft with and without alternative (emergency) landing option

When an alternative landing site exists, the same rotorcraft will perform a forward climb-out, following a small vertical climb, so that if the flight has to be aborted, it may be able to land safely. In any case, it must be able to demonstrate the same capability, e.g. clearing a minimum height of 35 ft with one engine inoperative in unfavourable conditions.

4.6   Emergency Landing Operations

Emergency landing operations occur for a variety of reasons, which include on-board emergencies, a fire, or a minor mechanical failure. In this instance we discuss the consequences of losing engine power, partially or totally. On a twin-engine rotorcraft, one engine failure is compensated by the transmission system, and the remaining engine is required to meet an increased torque demand. Aside from the control issues around the actual process of disengaging one engine from the main power train Chap. 3, there are flight mechanics issues and flight control response time.

In case of total loss of engine power, with any type of rotorcraft configuration, the default recovery strategy relies on using autorotation to slow down the loss of rotor rpm and the loss of flight altitude. *For a given gross weight and flight altitude, recovery and landing in autorotation depends critically on the relationship between flight altitude and airspeed.*

FAR Part 27, concerned with light helicopters ($W < 2,741$ kg), establishes performance criteria in autorotation. In §27.87, it is reported that *if there is h-V combination for which landing is not possible with engine failure, the flight envelope of unsafe h-V combination must be determined.*

Thus, the onus is on the helicopter design authority to demonstrate such control requirements. A typical *h-V* chart looks like the one in Fig. 4.5.

In this chart we display two shaded areas, one in low speed and the other in high speed, where helicopter recovery in autorotation is *highly unlikely* ("Avoid"). In fact, in the most up-to-date versions of this chart we have confidence levels that depend on the reaction time of the pilot. This time is measured in seconds, with 1 s being rapid, and 3 s being dangerously slow. The shorter the response time, the wider is the chart in the V-axis. Also shown is an ideal take-off flight path, denoted by a thick blue line that avoids both danger areas in the diagram: this is the safe take-off trajectory at the safe take-off speed, V-Toss.

Recovery and control is more problematic in hover. At the lowest point of the chart, with $V = 0$, there is a sudden vertical descent which can only be slowed down by quick pitch control and and aircraft flare to soften a crash landing. At the highest point, there is the possibility of gaining some horizontal speed before attempting to regain control. In any case, the entry point into autorotation is critical. As pointed out by Prouty [1], *a bad autorotation is usually survivable, but a bad beginning of an autorotation is usually not.*

The analysis of autorotational performance begins with consideration of the autorotative index, defined as the kinetic energy stored in the rotor mathematically, there are several definition. The physical definition is $AI = I_b \Omega^2 / 2W$ (I_b is the polar moment of inertia of the blade); in practice some engineers prefer to use a practical definition:

$$AI = \frac{I_b \Omega^2}{2\, W D_L} \tag{4.4}$$

where D_L is the disk loading, and the numerical value expressed in [ft³/lb]. Light helicopters like the Bell 206 and the Robinson R-22 have a high autorotation index ($AI \simeq 35-40$ ft³/lb, or $2.2-2.5$ m³/kg), and heavy-lift helicopters, such as the CH-53E have a relatively low AI ($AI \simeq 10$ ft³/lb), which makes an emergency landing more problematic. However, the CH-53E is a three-engined rotorcraft, rather than a single engine, and the odds of total engine failure are considerably lower.

By defining a critical height h_{cr} and a critical speed V_{cr}, Pegg [2] demonstrated that it is possible to correlate test points corresponding to a variety of gross weights and flight altitudes into a single curve. The critical speed is the largest speed in the low-speed danger area in Fig. 4.5, and h_{cr} is its corresponding flight altitude. The normalised plot is V/V_{cr} versus h/h_{cr}. One such chart is shown in Fig. 4.6 for selected flight data points.

Within an autorotative landing operation, there are five distinct phases.

The problem of autorotation is a subject that has vast coverage in the technical literature, from the very beginning [3, 4] to this day. The most recent research demonstrates that physiological aspects, pilot training and workloads are important in the determination of the final outcome of an autorotative manoeuvre.

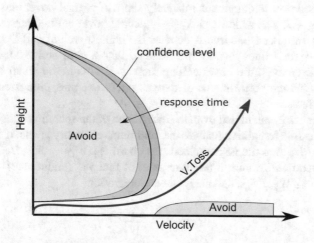

Fig. 4.5 Height-velocity chart

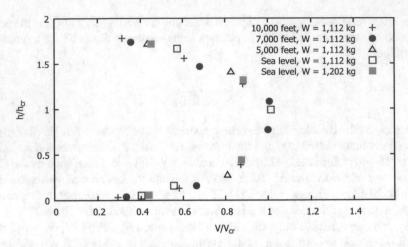

Fig. 4.6 Height-velocity chart in normalised format. Data from Ref. [2]

4.7 Vertical Flight Performance

Hover is unique capability of the helicopter, but it is performed sparingly by most helicopters. Only a few helicopters specialise in low-speed and hover operations, such as the Kaman K-1200 K K-Max, which is a single-engine (Honeywell T53-17A-1) synchropter with high-weight sling load capability.

There are safety issues, as discussed in the previous section, and fuel burn rates are high; thus, hover endurance is relatively short. A typical hover chart looks like the one in Fig. 4.7. The solid lines A, B $\cdots$ denote hover ceiling curves at constant gross weight. In this case, curve A denotes the highest weight and D is the lowest weight. The dashed lines display the standard day, a cold and a hot day. As the temperature decreases, the hover ceiling increases. This is the result of improved aero-thermodynamic performance of the gas turbine engine, alongside favourable air density effects.

The endurance is calculated from the integration of the specific endurance, up to a point when the fuel remaining reaches the minimum regulatory levels. If the fuel burn rate is dW_{f6}/dt, then the instantaneous fuel burn is $dW_{f6} = W_{f6}dt$. To calculate the time T_e required to burn a target amount of fuel W_f^* (endurance), we take the time step $dt = dW_{f6}/W_{f6}$ and carry out an integration

$$T_e = \int_o^{W_f^*} \frac{dW_{f6}}{W_{f6}} \tag{4.5}$$

At each point we need the fuel flow, which is calculated by a complete power analysis of a trimmed rotorcraft in hover. Note that in many cases the endurance quoted in the *Flight Crew Operating Manuals* refers to a loiter speed rather than hover.

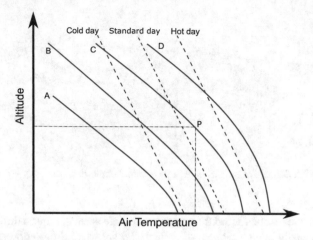

Fig. 4.7 Typical hover chart

4.8 Climb Performance

In Sect. 4.5 we discussed climb-out procedures. Now we calculate the power required to climb. This is done using *excess power*, defined as

$$v_c = \frac{P_e - (P_{mr} + P_{tr} + P_a + P_{gbox} + \cdots)}{W} \tag{4.6}$$

where the total power depends on the rotorcraft configuration. In this case, we have indicated the presence of a tail rotor. The gross weight requires a further clarification. In high-powered climb, the rotor wakes have a low skew angle and flow around the airframe, which may have a considerable amount of blockage, for example due to landing gear, sponsons, external stores, etc. This blockage generates a so-called *vertical drag* that must be overcome along with the gross weight. The net result is an increase in *apparent* weight. Although the inboard sections of the blades do not produce much thrust, there is some interference that must be accounted for, and in some heavy rotorcraft this additional download can be of the order of 7−12%. Figure 4.8 shows an example of a heavy-lift helicopter in such a situation. The wake flows around the sponsons and may interfere with the tail-rotor inflow, point B.

A low-order method for the estimation of this vertical drag is shown by Stepniewsky & Keys [5], and is based on using the strip theory. The result is that the gross weight is corrected as

$$W_1 = W + q A_1 C_{DV} \tag{4.7}$$

where q is the dynamic pressure based on the vertical flow, A_1 is the wake blockage area, and C_{DV} is a vertical drag coefficient. Based on flat plate theory, $C_{DV} \sim 1$.

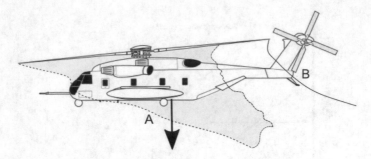

Fig. 4.8 Wake interference of the CH53 in high-powered climb-out.

The climb rate reaches a maximum a intermediate speeds, around the loiter speed. This result is easily justified, because at the loiter speed minimum power is required to fly at constant altitude, and more excess power is available for climb. At the maximum speed, there is no climb margin.

4.9 Ground Effects

The wakes considered so far are free from external constraints, but in reality when the rotorcraft operates near the ground there is strong interference, due to the fact that the wakes cannot penetrate solid walls and are thus forced to bounce back and spread outwards. There is a variety of situations, out of which we select a few for discussion. The simplest case is a rotor in hover near the ground. It has been demonstrated that at constant shaft power, the required hover thrust decreases. Vice versa, at a fixed thrust, the required power decreases. Ground effect only materialises when the rotor has a ground clearance of one rotor diameter or less. The more widespread empirical correlations are those of Cheeseman and Bennett [6] and Hayden [7], the latter one based on more recent ground-effect tests. The former equation for ground effect in hover is written as

$$\left(\frac{T_{oge}}{T_{ige}}\right)_P = \frac{1}{1 - (\bar{z}/4)^2}, \qquad \left(\frac{P_{oge}}{P_{ige}}\right)_T = \frac{1}{A + B\bar{z}^2} \qquad (4.8)$$

where $\bar{z} = z/R$ is the ground clearance normalised with the rotor radius, A and B are appropriate empirical coefficients.

When the rotor advances in ground effect, the wake takes complicated shapes that depend on the rotor advance ratio. An extended theory, also due to Cheeseman, indicates that at advance ratios $\mu \geq 0.1$ the ground effect is effectively negligible, because the wake is rapidly convected downstream past the aircraft. The extension of this theory yields the following:

$$\frac{T_{oge}}{T_{ige}} = \frac{1}{1 - (\bar{z}/4)/(1 + \tan^2 \chi)} \qquad (4.9)$$

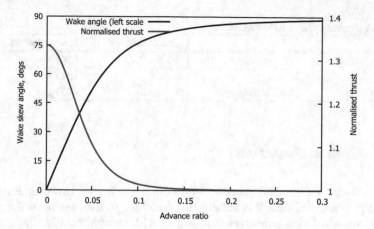

Fig. 4.9 Wake skew angle in ground effect, $C_T = 0.01, \bar{z} = 0.5$.

where χ is the wake skew angle. This skew angle is related to the advance ratio μ and the mean inflow velocity ratio λ_i

$$\cos \chi = \frac{\mu}{\lambda_i} = \frac{2}{C_T} \left(\sqrt{\mu^4 + C_T^2/4} - \mu^2 \right) \tag{4.10}$$

displayed in Fig. 4.9, which shows no practical effects at $\mu > 0.15$. It is noted that the result of Eq. 4.10 depends on the C_T. A normalised advance ratio is defined as $\mu^* = \mu \sqrt{2/C_T}$.

Experimental data have been published, for example Ref. [8], which provide evidence two distinct flow regimes, at low- and high rotor advance ratios. At normalised advance ratios $\mu^* > 1$ the ground vortex below the rotor disappears, and so does the ground effect. This event corresponds to $\mu \simeq 0.007$, Fig. 4.9.

Another important ground effect problem occurs when operating to/from unprepared surfaces, which means there is loose ground below the aircraft. In this instance we have examples of brownout, white-out and a plethora of other problems caused by strong downwash.

The brownout problem arises when the rotorcraft attempts to land onto very loose ground, causing the raise of dust and other particulate that generate a large bowl of cloud enveloping the complete aircraft and obscuring the field of view of the pilots.

The disk loading is limited by ground operations. In fact, large disk loadings correspond to large downwash velocities, which may create hazards for personnel on the ground. Some practical limits are given in Table 4.1.

Table 4.1 Summary of disk loading limits

		Firm ground	Loose ground
Personnel limits	Over-turning moment	400 Nm	300 Nm
	Force	450 N	330 N
Surface limits	Disk loading	245 kg/m^2	73 kg/m^2

4.9.1 Ground Operations

Rotorcraft with landing skids can only touch down and lift off. However, if wheeled landing gear are available, it is possible (in principle) to manoeuvre on the ground at small speeds. In extreme cases, helicopters with conventional landing gear can using their ground rolling capability to take off from airfields at altitudes above their certified altitude. In fact, ground rolling takes the rotorcraft to a speed that corresponds to a lower forward flight power. The taxi speed is found from

$$kT_{ige} \sin \alpha_r = D + \mu_r \mathcal{R} \tag{4.11}$$

where α_r is the rotor disk tilt angle, T_{ige} is the thrust in ground effect, k is a blockage factor, due to fuselage and undercarriage interference, D is the aerodynamic drag of the rotorcraft, μ_r is a rolling coefficient depending on the runway conditions, and $\mathcal{R} = W - L$ is the rolling resistance. By expanding all the terms, Eq. 4.11 becomes

$$kT_{ige} \sin \alpha_r = \frac{1}{2}\rho A C_{D_g} U^2 + \mu_r \left(W - kT_{ige} \cos \alpha_r\right) \tag{4.12}$$

In Eq. 4.12, A is the rotor disk area, used as a reference for the drag coefficient in ground effect, C_{D_g}. The rotor downwash has a limit, usually specified so that there is no harm to ground personnel. Data from Table 4.1 can be used. Actually, by specifying the limit downwash (or limit disk loading), we can calculate the limit thrust in ground effect, and hence the limit taxi speed. Now divide Eq. 4.12 by the disk and solve for the ground speed V

$$V^2 \simeq \frac{2}{\rho C_{Dg}} \left[k\frac{T_{ige}}{A} \sin \alpha_r - \mu_r \left(\frac{W}{A} - k\frac{T_{ige}}{A} \cos \alpha_r\right)\right] \tag{4.13}$$

This expression contains the nominal disk loading W/A and the equivalent disk loading in ground effect T_{IGE}/A, for which we need to set a limit $T_{A_{max}}$. Note that the vehicle is capable of taxiing only if

$$W < \frac{k}{\mu}(T/A)_{max} \left(\sin \alpha_r + \cos \alpha_r\right) \tag{4.14}$$

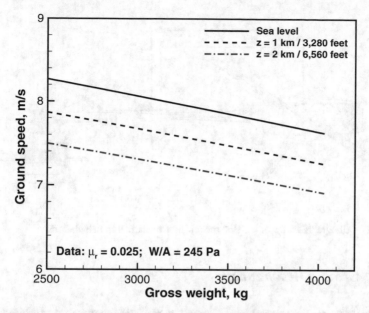

Fig. 4.10 Calculated ground speed of reference helicopter at selected altitudes

Since is the speed is generally very small, changes in thrust with the speed can be neglected. The thrust T_{IGE} will be calculated from the condition of constant power derived by Cheeseman and Bennett [6]

$$\left(\frac{T_{ige}}{T_{oge}}\right)_P = \frac{1}{k_g} \tag{4.15}$$

with k_g the ground effect factor,

$$k_g = 1 - \frac{(R/4z)^2}{1 + (\mu/\lambda_i)^2} \tag{4.16}$$

having neglected the effects of blade loading on the rotor; In Eq. 4.16 h is the height of the hub on the ground; $z/2R$ is the ground clearance of the rotor.

A solution of Eq. 4.13 is shown in Fig. 4.10. The problem's parameters are: $\mu = 0.025$ (dry hard ground); $W/A = 245$ Pa (limit rotor downwash), or $P = 100$ kW.

An interesting case is that of the autogyro. Since this rotorcraft cannot hover and fly vertically, take-off and landing must be achieved through a ground run. Therefore, this type of rotorcraft *must have wheeled landing gear* in all cases.

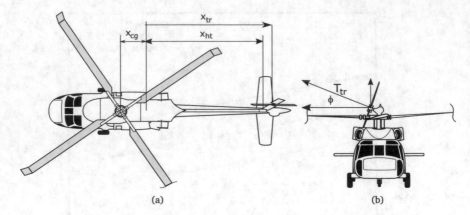

Fig. 4.11 Orthographic view of a UH60 model for trim static trim calculations

4.9.2 Ground Resonance

In the early days of rotorcraft development, some helicopter rotors exhibited violent vibrations when on the ground. In some cases these vibrations were disastrous. At first this problem was attributed to rotor blades flutter, but later it was understood to be caused by a hitherto new phenomenon: the conversion of rotational energy into vibration energy in the presence of the ground. Hence the phenomenon was called *ground resonance*. The problem was solved mathematically by Coleman and Feingold [9] in the 1940s, but it is quite interesting to this day to observe video recordings of rotorcraft running to self-destruction in some ground tests.

Ground resonance is an instability of the rotor that is placed on a flexible surface. An imbalance or perturbation on a rotor blade causes a forcing to be transferred to the airframe. The airframe is on the ground standing on flexible landing gear The perturbation can be transferred to the landing gear and back (amplified) to the airframe, and onward back to the rotor. When these oscillations self-amplify, there is resonance.

4.10 Static Trim Conditions

A helicopter is said to be trimmed if all forces and moments are balanced, and the helicopter can fly steady-state level flight. The solution of the trim equations, easy in principle, requires considerable mathematical treatment, and in most cases is requires numerical solutions. The terms in trim equations depend on the specific rotorcraft.

Consider a helicopter that has a tail rotor with a cant angle φ, a horizontal stabiliser, such as the UH60. The vertical trim and the pitching moment are written

$$T \cos \alpha_r + T_{tr} \cos \varphi + L_{ht} = W$$

$$x_{cg} T \cos \alpha_r - x_{tr} T_{tr} \cos \varphi - x_{ht} L_{ht} = 0 \tag{4.17}$$

where α_r is the forward tilt angle of the main rotor, L_{ht} the stabiliser lift and the x-distances are indicated in Fig. 4.11. There are too many unknowns in this equation to be solved, and thus we need further elaboration. The stabiliser lift is

$$L_{ht} = q \, (A C_{L\alpha} \alpha)_{ht} \tag{4.18}$$

and requires the effective inflow angle α_{ht}, which may be dissimilar between the two sides of the stabiliser, due to the rotor downwash and the presence of the fin; q is the dynamic pressure based on the flight speed, A_{ht} is the area of the stabiliser used to calculate the aerodynamic derivative $C_{L\alpha ht}$. The tail rotor thrust depends on the yaw trim condition, and is written

$$T_{tr} \sin \varphi = \frac{1}{x_{cg} + x_{tr}} \frac{P_{mr}}{\Omega} \tag{4.19}$$

The main tilt angle of the thrust α_r depends on the total drag of the helicopter; thus, we require a trim equation in the horizontal direction

$$T \sin \alpha_r = D \tag{4.20}$$

unless the tail rotor also has a tilt in the forward direction that provides a contribution to the total thrust (now excluded to avoid further complications). The helicopter drag is unknown, and can only be estimated, as it depends on the airframe and its inflow/yaw angles, the rotor head, the rotor systems and any ancillary elements.

Hence it is clear that the equations, although algebraically simple when written on their own, suddenly become complicated because of the cross-correlation effects.

The next case is that of a helicopter with an asymmetric empennage, in particular two vertical stabilisers at a yaw angle with respect to the longitudinal axis of the aircraft, such as the Eurocopter (now Airbus) AS365, Fig. 4.12.

The lateral trim equation is

$$Q_{mr} + Q_{tr} + Q_{vt} = 0 \tag{4.21}$$

where we have included the effects of the vertical stabilisers. These contribute a side force Y_{vt}, which depends on the design yaw angle β and the flight yaw angle β_∞:

$$Y_{vt} = q A_{ref} C_{y\beta} (\beta + \beta_\infty), \qquad Q_{vt} = x_{vt} Y_{vt} \tag{4.22}$$

where $C_{y\beta}$ is the side force derivative of the stabiliser with respect to the yaw angle β; A_{ref} is a reference area. This coefficient is calculated with aerodynamic considerations and a separate demonstration is given through a *video tutorial*.

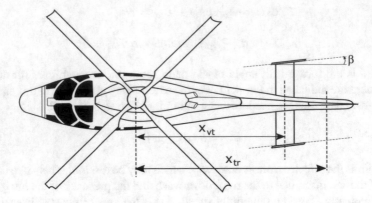

Fig. 4.12 Top view of an AS365 model for torque balance calculations. This helicopter has a fenestron, which is not visible in this graph

Assume $\beta_\infty = 0$ to simplify the analysis. The fenestron thrust required is

$$T_{tr} = \frac{1}{x_{tr}} \frac{P_{mr}}{\Omega} - q A_{ref} C_{y\beta} \beta \qquad (4.23)$$

which demonstrates that with a well-designed stabiliser, *for some flight conditions*, the stabiliser contributes to the overall side force and thus offsets the requirements on the tail rotor power.

For more general analyses, we need a variety of stability derivatives, as demonstrated in earlier literature [10–12]. Modern examples are shown in Chap. 5 for the convertible rotor.

4.11 Helicopter Speed Stability

The speed stability problem is related to the change of pitching moment with increasing speed. The helicopter can have an inherent nose-up or nose-down pitching moment. With reference to Fig. 4.13, using the control stick fixed, if the rotorcraft has a nose-up C_M, it will decelerate, because its thrust line tilts backward. If the rotorcraft has nose-down C_M, it accelerates further, because of its nose-down attitude. The nose-down pitching moment is destabilising and the helicopter is said to be unstable with speed. A nose-up pitching moment is stabilising. The flight control system must be able to compensate this pitch tendency.

A particularly interesting case is that of the tandem helicopter, a case studied by Tapscott and Amer [13] who provided a simplified analysis for the collective differential required to maintain longitudinal stability of this type of helicopter. The key assumptions of this model include equal-diameter rotors, equal rotor solidity, and equal angular speed. Critical to the whole procedure is that the fore rotor is

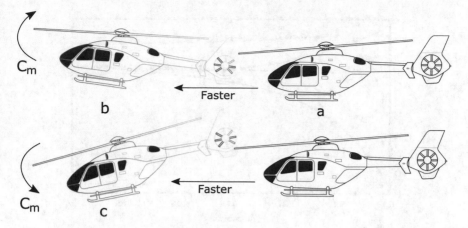

Fig. 4.13 Speed stability of a conventional helicopter

unaffected by the aft rotor; therefore, the aerodynamic interference is neglected, just as the contributions of the airframe lift and pitching moment. With these assumptions, it is possible to show that the collective differential between the two rotors (or swashplates) is

$$\frac{\Delta T}{W} = \frac{1}{2} \frac{1}{C_{T\sigma}} \left[\left(\frac{\partial C_{T\sigma}}{\partial \alpha} \right) \Delta\alpha + \left(\frac{\partial C_{T\sigma}}{\partial \theta} \right) \Delta\theta \right] \qquad (4.24)$$

where ΔT is the thrust differential, $\Delta\theta$ is the difference in collective pitch, $\Delta\alpha$ is the difference in mean inflow angle on the rotors and $C_{T\sigma} = C_T/\sigma$ is the mean blade loading. The derivative are mean values between the two rotors. Equation 4.24 must be solved for $\Delta\theta$. To begin with, we differentiate with respect to the advance ratio and set $\partial\Delta T/\partial\mu = 0$ to eventually find the following expression:

$$\frac{\partial \Delta\theta}{\partial\mu} \simeq k_1 \overline{C}_{T\sigma} \frac{\Delta T}{W} + k_2 f(\Delta\sigma, \Delta V_{tip}) + k_3 \Delta\alpha_d + k_4 \overline{C}_T \qquad (4.25)$$

where k_1, k_2, k_3, k_4 are stability derivatives depending on the blade loading and the advance ratio; $\Delta\alpha_d$ is a swashplate *dihedral effect*, one effect that is difficult to estimate, but discussed in the original work. The second term with stability factor k_2 can be neglected if there is no difference in tip speed and rotor solidity.

A plot of these functions is shown in Fig. 4.14. When applied to a model of the CH-47 helicopter, the collective differential increases rapidly at high speeds from a negative value. The predicted collective differential $\Delta\theta$ is displayed in Fig. 4.15.

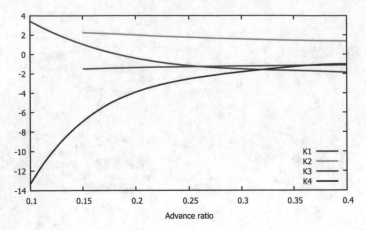

Fig. 4.14 Stability coefficients in Eq. 4.25

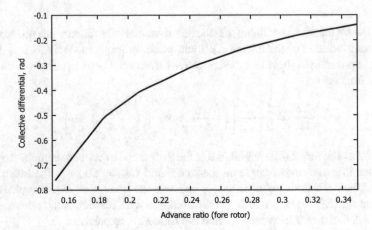

Fig. 4.15 Predicted collective differential for the CH-47

4.12 Mission Planning

Mission planning is the series of calculations of fuel requirements for a specified operation. In the process, we define the take-off gross weights and analyse the mission limitations, depending on required payload, atmospheric conditions and other external factors.

The process begins with the identification of the flight segments involved in the forecast mission. This is sometimes straightforward (flight from A to B to deliver a payload), but in many cases is a rather complicated undertaking that involves scenario forecasting, risk analysis and contingency planning. This is particularly true for search-and-rescue and most military operations. In these cases, we might not have an exact definition of flight segments, but we need to plan around an extension of the flight, the duration of a hover and refuelling, on the ground or airborne.

In the simplest case, we have a full list of flight segments, we calculate the fuel required for each segment, sum all these fuel contributions, add a contingency amount of fuel to reach a first-order estimate. The correct process is as follows:

- Establish the required payload, W_p.
- Establish the atmospheric conditions.
- Establish the flight segments, S_i.
- Make an initial guess of the gross take-off weight, $W_{j=1}$ (j is the iteration count).
- Calculate the fuel required for each segment, W_{fi}, each corresponding to S_i.
- Calculate the regulatory fuel reserves, W_{fr}.
- Calculate the sum of all fuel required: $W_f = \sum_i W_{fi} + W_{fr}$.
- Calculate the gross take-off weight: $W_{j+1} = W_e + W_p + W_f$.
- Calculate the relative difference in gross take-off weight: $\epsilon = |W_{j+1} - W_j|/W_j$.
- Establish a convergence criterion, for example $\epsilon < 10^3$.
- if $\epsilon > 10^3$, restart the process with the new gross take-off weight W_{j+1} until convergence.

In most cases, convergence is achieved in a few iterations, but in a very few instances, this numerical method does not converge. Upon convergence, it is possible that the final weight $W >$ MTOW, in which case, the mission cannot be fulfilled: the mission specification must be changed and/or the payload must be reduced.

Each flight segment must be specified by a few parameters which include: flight time, initial and final altitudes, initial and final air speed, climb rate (mean value), other specific data. Examples of mission calculations are shown below.

Mission Scenario #1: Medevac operation of a CH-47. We allow the rotorcraft to land, after search operations, slow down the rotors, load as many casualties as possible, take-off and start the return journey, with the listing of flight segments shown below.

A warm-up of 3 min at an altitude of 80 m (~260 ft) is followed by a climb-out to a target altitude of 305 m (~1000 ft), to a target air speed of 100 kt, with an average climb rate of 3 m/s (~600 ft/min). Then there is a cruise at 100 kt, at 305 m altitude to a target distance $X = 100$ n-miles. Loiter for search-and-rescue for up to 10 min at 1000 ft (305 m), then descend and land to a target altitude of 100 m (328 ft), and so on.

Note that in this case there is a segment called "weight-load", after landing, requiring the rotorcraft to load up to 3000 kg in 15 min (200 kg/min); then the rotorcraft makes the return journey.

The complete analysis is demonstrated in a *video tutorial*, alongside a number of variations on the medevac operation, including a case when landing is not possible and loading of casualties has to be done whilst airborne; there is also another case when both landing and refuelling are possible.

Listing 4.1 *Example of medevac operation.*

# FLIGHT	time [min]	z [m]	V [kt]	X [nm]	Other [−]	Notes
#						
#						
'Medical Evacuation'						
'Payload' 100						! initial payload weight
'warm–up'	3	80	0	0	0	! z = airfield altitude
'climb–out'	0	305	100	0	3	! climb rate (m/s)
'cruise'	0	305	100	150	0	! z = cruise altitude; V > 0
'loiter'	10	305	0	0	0	
'descent'	0	305	0	0	0	! z = initial descent altitude
'landing'	0	80	0	0	0	! z = airfield altitude
'hover'	0.5	10	0	0	0	! cargo loaded in hover
'weight–load'	15	10	0	0	3000	! cargo loaded on the ground
'climb–out'	0	305	100	0	6	! climb rate (m/s)
'cruise'	0	305	100	150	0	! z = cruise altitude; V > 0
'descent'	0	305	0	0	0	! z = initial descent altitude
'landing'	0	100	0	0	0	! z = airfield altitude
'END'						! END PARSING OF DATA

4.13 Payload-Range

The most common way of demonstrating a mission performance is via a payload-range diagram. This is also the simplest possible mission of a rotorcraft, since it implies a flight from origin to destination with a fixed payload. Whilst this is the appropriate performance description of a fixed-wing aircraft, it is rather limiting for a rotorcraft, as a result of the peculiar flight characteristics of this vehicle. In any case, payload-range diagrams are produced by manufacturers to demonstrate range and payload capability as part of their overall marketing strategies.

A typical payload-range diagram appears as in Fig. 4.16. In graph 4.16a, the segment A-C indicates that the payload decreases as the amount of fuel increases, subject to the aircraft starting at MTOW. Upon reaching range X_C, any further increase in range can only be achieved by dropping payload, subject to the rotorcraft starting with full tanks. The gross take-off weight in the segment C-D is lower than the MTOW, and upon reaching distance X_D there is virtually no payload capacity left. Point D is called *ferry range*. The case of Graph 4.16b corresponds to a case when the rotorcraft has been equipped with additional fuel tanks. For any flight range $X < X_C$ because the additional tanks add to the structural empty weight of the vehicle. This is inevitable, but it also indicates that additional tanks are only to be fitted on rotorcraft specifically allocated to long-range or long-endurance missions.

As demonstrated in the case of the medevac operation, Sect. 4.12, performance charts like the ones in Fig. 4.16 do not reveal the whole performance envelope of the vertical lift vehicle; therefore, alternative charts have to be used. For example, an alternative performance charts would account for a specified amount of time in hover and loiter. This amount of time can be somewhat arbitrary, but for the sake of discussion, let us consider the case of a rotorcraft requiring a minimum 10 min hover alongside 10 min of loiter for search-and-rescue. If we introduce such considerations in the flight operations, we still face some arbitrary decisions. For example, at what point into the flight is the rotorcraft required to hover and/or loiter?

4.14 Direct Operating Costs

Ownership and operation of helicopters is notoriously expensive, and the costs of ownership are possibly the highest barrier to entry for many potential customers, alongside flight safety. A good costs analysis highlighting the difficulty with helicopters is shown in Ref. [14]. In that work, it was demonstrated how acquisition costs are critically dependent on installed power, rather than empty weight. This result would translate into a relatively higher cost for a rotorcraft with a high design disk loading, and it was concluded that *designing for minimum empty weight does not equate to minimum helicopter purchase cost.*

In this section, we discuss the overall costs, which include – critically—the direct operating costs, beginning with ownership models. There are several ownership models; often the operators are not aircraft owners, but rather leasers of the vehicles. This poses further questions as to the structure of the costs. *Direct Operating Costs (DOC)* are the sum of all costs that are incurred by an owner/operator. Ultimately, it all comes down to a very important economic figure: DOC/flight hour.

A summary of cost items is provided in 4.2. In this case, we have three types of operations: transport, training and heavy lifting. For each type of operation, the data to inser in the table must be *pre-calculated*, and are intended as averages, otherwise

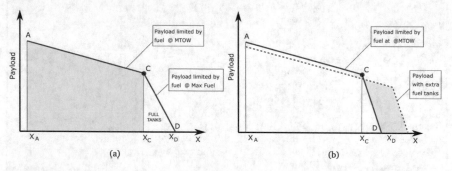

Fig. 4.16 Payload range charts of a helicopter

we would need a mission calculation for each specific mission. Thus, the data that must be pre-calculated include: block time and fuel burn.

Listing 4.2 Summary of Direct Operating Costs

#	Item	Unit
1	Item	Unit
2		
3	aircraft price	any unit (dollars)
4	fuel price	/kg
5	fuel price inflation, estimated	%
6	insurance, based on aircraft actual value	%/year
7		
8	spares price1 (airframe, landing gear, tyres)	/flight hour
9	spares price2 (engine, APU, lubricants)	/flight hour
10	spares price3 (avionics/systems/pax_services)	/flight hour
11		
12	life time	
13	depreciation over life–time	% of initial cost
14	financing	% of initial cost
15	interest rate	%
16	years of repayment	
17		
18	crew price, pilots	full time/pilot/year
19	crew price, ground	full time/staff/year
20	crew price inflation	% year
21	off station price	/person/night
22	labour rate1, engines	man–hour
23	labour rate2, in–house maintainance	man–hour
24	labour rate3, contracted out	man–hour
25	man–hour1, power plant	/flight hour
26	man–hour2, in–house	/flight hour
27	man–hour3, contracted out	/flight hour
28	landing charges, airfield services	/operation
29	ground handling costs	/movement
30	recurrent training costs	crew training/year
31	ground service costs: hangars	/year
32		
33	transport	type of operation
34	300	cycles in category this category	cycles/year
35	1500.	fuel burn in mission transport	[kg]
36	80.	block time	[min]
37	220.	stage length, n–miles (transport)	[average]
38	3000.	cargo/freight (inside)	[kg]
39	4000.	cargo/freight (under–sling)	[kg]
40	5.25	cargo price (inside)	[kg]
41	7.50	cargo price (under–sling)	[kg]
42	1.0	any other cost	/flight
43		
44	training	type of operation
45	10	cycles in category this category	cycles/year
46	500.	fuel burn in full mission training	[kg]
47	60.	block time	[min]
48	3000.	cargo/freight (inside)	[kg]
49	4000.	cargo/freight (under–sling)	[kg]

50	5.25	cargo price (inside)	(currency/kg)
51	7.50	cargo price (under–sling)	(currency/kg)
52			
53	heavy-lift	type of operation	
54	10	cycles in category this category	cycles/year
55	1000.	fuel burn in full mission heavy–lift	[kg]
56	60.	block time	[min]

The use of this configuration table is demonstrated in a *video tutorial*.

4.15   Performance Augmentation

One of the most-often heard criticisms of the helicopter is that it is not fast, and for quite some time engineers have worked on concepts to increase the speed by using *compounding*. The criticism is unwarranted, because the helicopter can carry out missions that are not comparable with any fixed wing – just think of the ability to deliver large external loads, almost anywhere, with high precision.

A flight envelope is a safe operating domain in the z-V space. Generally, it refers to steady state flight, but accelerations and manoeuvres are possible in this domain. An expansion of the flight envelope requires higher speed and higher flight altitudes.

Let us start with the flight speed. Speed is good, but not at all costs. The *Westland Lynx* that achieved a world speed record [15] was an exception in that it was a conventional helicopter finely tuned for speed, but that was not a production version.

Compounds are meant to increase speed. There are two basic type: *thrust compounds*, that increase the net thrust via additional thrusters, which can be propellers (ducted or unducted), or jet engines. Examples in this category include the *Piasecki X-49* (with aft ducted propeller), the *Sikorsky X2* and *S-97* (coaxial rotor with aft pusher propeller), the *Sikorsky S-69* (coaxial with two auxiliary side-by-side turbojets), and the *Airbus X-3* (with two wing-mounted left-right propellers).

Then there is the *lift compound* which is capable of generating additional lift through lifting surfaces; at the same time they offload the rotor, which can be slowed down to limit retreating blade stall and high-Mach flows on the advancing blades (see also Fig. 4.2). Examples in this category include the *Sikorsky S-67* and the *Lockheed AH-56*.

This is a vast subject that needs its own chapter, but a number of references can help to start out [16, 17]. Only a few concepts are given below. The thrust compound helicopter requires a separate power plant or engines that are oversized to drive at least one more propulsor; they are the domain of military operations.

Another limiter in helicopter flight performance is the relatively low operational altitude limits. These are essentially due a combination of sharp increase in induced power ($P_i \sim 1/\sqrt{\rho}$) with a rapid decrease in net turboshaft power at high altitudes. Although the rotors can be optimised, and the weight can be reduced, the only way to fly higher is to increase the installed engine power, as explained graphically in

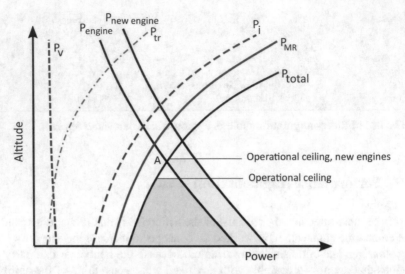

Fig. 4.17 Analysis of operational ceiling of a conventional helicopter; "A" denotes the absolute ceiling with the standard engine

Fig. 4.17. By further decreasing the gross weight, the required power shifts to the left, and the power margin increases, which allows the rotorcraft to fly higher.

For example, the *Eurocopter/Airbus AS350-B3*, powered by a *new* Turbomeca Arriel 2B turboshaft engine rated at 632 kW[1], would have an operational ceiling of 3415 m (11,200 ft) on a standard day and MTOW. With the weight reduced by 650 kg (by dumping payload), the ceiling increases to 6520 m (21,390 ft). By relaxing the constraint on the minimum climb rate to 1 m/s (200 ft/min), this lighter weight configuration can climb and operate at 7000 m (22,965 ft). Further engine upgrades, and a cold day, can take this helicopter up to mount Everest.

4.15.1 Lift Compound

Lift compounds have the problem of aerodynamic interference between the rotor downwash and the wing, which may cause off-design inflow conditions (large and inflow angles on the wing, asymmetric dynamic pressure, vertical drag, etc.), which makes them unsuitable for some flight conditions. However, they do not require an additional propulsor, which has some advantages in terms of complexity, weight and cost. Size and position of these surfaces is critical, because as mentioned they can cause interference in vertical- and low speed flight. Ideally, the wing would be tiltable and retractable, but there are design complications and costs to consider.

---

[1] Data from *The Flight Crew Operating Manual*.

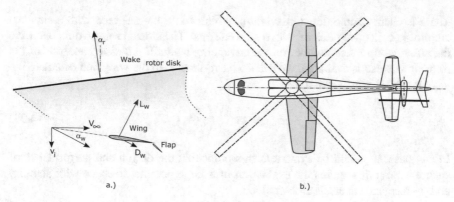

Fig. 4.18 Lift compound helicopter model

Now assume the rotor-wing combination shown in Fig. 4.18. The wing is just
below the rotor. Unless the wake is highly skewed to avoid the wing (Fig. 4.9), as it
happens in fast forward flight, the aerodynamic interference is too strong.

Due to the asymmetry of the rotor flow, each semi-wing may generate its own lift,
which causes additional rolling/yaw moments on the aircraft. Thus, one may want to
consider asymmetric wings as well, but there is no guarantee that these moments will
be automatically trimmed at all flight conditions. To secure more control authority,
we may need additional surfaces, such as vertical and horizontal stabilisers with
elevators and rudders. The lift generated by the rotor-plus-wing configuration in
steady level flight is

$$W = T \cos \alpha_r + (L_{w1} \cos \alpha_{w1} + D_{w1} \sin \alpha_{w1}) + (L_{w2} \cos \alpha_{w2} + D_{w2} \sin \alpha_{w2})$$
(4.26)

where L_w and D_w are the total wing lift and drag, respectively; the two numbers
indicate the wing on either side of the fuselage. If the wing is fixed, the inflow angle
α_w depends on the rotor downwash, which in turns depends on the airspeed. The
effective inflow on the wing is the angle between the nominal chord line and the
direction of the vector $\mathbf{V} = \mathbf{v}_i + \mathbf{V}_\infty$, Fig. 4.18a.

To begin with, assume that the difference $\alpha_{w1} - \alpha_{w2}$ is negligible. In this case the
rotor thrust required (in absence of fuselage and tail rotor contribution) is

$$T^* = \frac{W - L_w \cos \alpha_w + D_w \sin \alpha_w}{\cos \alpha_r}$$
(4.27)

with

$$L_w = q C_{l,\alpha} (\alpha_w - \alpha_o) b_w \bar{c}_w$$
(4.28)

There is clearly some thrust alleviation, which could be delivered with a smaller-amplitude collective pitch or with a slowed rotor. The slowed rotor would decrease the effective tip Mach number on the advancing blade. If a detailed analysis of the rotor downwash is available, we can use strip theory on the wing and calculate the local inflow at wing section j;

$$\alpha_{wj} = \tan^{-1}\left(\frac{v_{ij}}{V_\infty}\right) \tag{4.29}$$

In any case, it should be evident at this point that the design and performance of such a rotorcraft requires the evaluation of a large parametric space with stability and performance analysis all combined.

4.15.2 Thrust Compound

Some of the difficulties highlighted at the previous points are overcome with a thrust compound vehicle. The vertical component of the thrust remains the same. Thus:

$$W = T\cos\alpha_r, \qquad T\sin\alpha_r + T_p = D \tag{4.30}$$

where T_p is the thrust generated by an aft-mounted propeller. The tilt angle of the rotor disk becomes

$$\tan\alpha_r = \frac{D - T_p}{W} \tag{4.31}$$

Unless $D > T_p$, the rotor would be tilting backward and the rotorcraft would cease to operate as we know it: *a backward tilt would essentially convert the compound helicopter to an autogyro.* This reversal condition provides the limits of the propeller thrust. In hover and low speeds, the propeller may need to be disengaged.

One drawback of this configuration, seldom advertised, is the very large noise created by the interaction of the aft rotor with the main rotor. Wake interference effects are almost inevitable, and the propeller ingests turbulence and vortices from the main rotor. Furthermore, the tonal noise components of the propeller are different from those of the main rotor and combine together to give a more complex acoustic spectrum. For example, the Sikorsky X2 has a rotor operating at 360 rpm in low speed ($V < 200$ kt), whilst the propeller rpm is 1400 (gear-ratio of 4.88); therefore, the blade passing frequency 24 Hz for the rotor 140 Hz for the six-bladed propeller (the frequency ratio is 5.83). Rotor acoustics is discussed in detail in Chap. 6.

One notable advantage of the Sikorsky X2 is that the use of counter-rotating coaxial rotors removes the need for a conventional tail rotor, and therefore there are only limited requirements on the torque balance, less tendency to yaw and roll. A coaxial rotor with rotors providing torque Q_1 and Q_2 is automatically stabilised in yaw only if $Q_1 = -Q_2$. With the lower rotor operating in the downwash of the upper rotor, a collective pitch differential is required.

References

1. Prouty RW (2009) Helicopter aerodynamics, vol I. Eagle Eye Solutions
2. Pegg RJ (1968) An investigation of the helicopter height-velocity diagram showing effects of density altitude and gross weight. Technical Report TN D-4536, NASA
3. Gessow A, Myers MG (1947) Flight tests of a helicopter in autorotation, including a comparison with theory. Technical Report TN-1267, NACA
4. Nikosky AA, Seckel E (1949) An analysis of the transition of a helicopter from hovering to steady autorotative vertical descent. Technical Report TN 1907, NACA
5. Stepniewsky WZ, Keys CN (1984) Rotary wing aerodynamics. Dover Editions
6. Cheeseman IC, Bennett WE (1955) The effect of the ground on a helicopter rotor. R & M 3021, ARC
7. Hayden JS (1976) The effect of the ground on helicopter hover power required. In: 32nd AHS annual forum proceedings, Washington, DC
8. Curtiss HC, Sun M, Putman WF, Hanker EJ (1984) Rotor aerodynamics in ground effect at low advance ratios. J Am Hel Soc 29(1):48–55. https://doi.org/10.4050/JAHS.29.48
9. Coleman R, Feingold AM (1958) Theory of self-excited mechanical oscillations of helicopter rotors with hinged blades. Technical Report 1351, NACA (Supersedes TN-3844 and TR 1351)
10. Prouty RW (1988) Military helicopter design technology. Krieger Publ Co (reprint)
11. Seckel E (1964) Stability and control of aircraft and helicopters. Addison-Wesley
12. Franklin J (2002) Dynamics. AIAA, control and flying qualities of V/STOL vehicles
13. Tapscott RJ, Amer KB (1956) Studies of the speed stability of a tandem helicopter in forward flight. Technical Report R-1260, NACA
14. Harris FD, Scully MP (1998) Rotorcraft cost too much. J Am Helicopter Soc 43(1):3–13
15. Perry FJ (1987) The aerodynamics of the world speed record. In: 43rd AHS annual forum proceedings, St Louis, MO
16. Erickson R, Kufeld R, Cross J, Hodge R, Ericson W, Carter R (1984) NASA rotor system research aircraft flight-test data report: Helicopter and compound configuration. Technical Report TM-85843, NASA
17. Yeo D (2019) Design and aeromechanics investigation of compound helicopters. Aerosp Sci Technol 88:158–173. https://doi.org/10.1016/j.ast.2019.03.010

Antonio Filippone is at the School of Engineering, University of Manchester, where he specializes in aircraft and rotorcraft performance, environmental emissions aircraft noise, and engine performance.

Chapter 5
Tiltrotor Aeromechanics

Wesley Appleton

Abstract This chapter explores several aspects of the tiltrotor configuration and covers challenges in their design, modelling and simulation. An overview of the aircraft is first given that details the operational role this configuration aims to fulfil and summaries the amalgamation of a rotary-wing and fixed-wing aircraft into a single flight vehicle. Two important challenges of the tiltrotor configuration are then examined: the interactional effects of the aircraft through different phases of flight; and the undetermined control problem arising from the existence of both rotary-wing and fixed-wing controls. The tiltrotor aeromechanics are then introduced to give an outline of the required modelling and simulation elements of the different aircraft components through their extensive flight envelope. Finally, a numerical example is presented to better understand the transitional regime of flight in terms of aircraft performance and trim behaviour.

Nomenclature

Roman Symbols

$A_{[\,]}$	Control input amplitude of subscripted quantity [rad]
b	Wing semi-span [m]
$C_{L_{\max}}$	Maximum coefficient of lift
C_T	Thrust coefficient, $C_T = T/\pi\rho\Omega^2 R^4$
C_P	Power coefficient, $C_P = P/\pi\rho\Omega^3 R^5$
D	Drag force [N]
e	Dimensionless flapping hinge offset
h	Altitude [m]
n	Thrust-to-weight ratio
P	Power [W]

Supplementary Information The online version contains supplementary material available at https://doi.org/10.1007/978-3-031-12437-2_5.

W. Appleton (✉)
School of Engineering, The University of Manchester, Manchester, UK
e-mail: wesley.appleton@manchester.ac.uk

© The Author(s), under exclusive license to Springer Nature Switzerland AG 2023
A. Filippone and G. Barakos (eds.), *Lecture Notes in Rotorcraft Engineering*,
Springer Aerospace Technology, https://doi.org/10.1007/978-3-031-12437-2_5

P_i	Induced power [W]
r	Radial ordinate [m]
R	Rotor radius [m]
T	Rotor thrust [N]
V_∞	Airspeed [m s^{-1}]

Greek Symbols

α	Angle-of-attack [rad]
β_{1c}	First harmonic of fore/aft rotor flapping [rad]
β, dβ/dψ	Blade flap angle and flap rate [rad], [-]
δ	Normalised fore/aft pilot stick position
η	Dimensionless radial ordinate, $\eta = r/R$
η	Elevator deflection [rad]
θ	Blade section pitch angle [rad]
θ_{tw}	Blade section twist angle [rad]
θ_0	Collective pitch [rad]
θ_{1s}, θ_{1c}	Longitudinal and lateral cyclic pitches [rad],[rad]
λ	Total inflow ratio
λ_i	Induced inflow ratio, $\lambda_i = v_i/\Omega R$
μ	Advance ratio
v_i	Induced velocity [m s^{-1}]
ρ	Air density [kg m^{-3}]
σ	Rotor solidity ratio
σ	Relative density
τ	Rotor tilt angle [rad]
ϕ	Blade section inflow angle [rad]
ψ	Blade azimuth angle [rad]
Ω	Rotor speed [rad s^{-1}]

Subscripts

E	Denotes the empennage
F	Denotes the fuselage
N	Denotes the nacelles
R	Denotes the rotors
TPP	Tip-path plane
W	Denotes the wing

Abbreviation

cg	Centre of gravity

5.1   The Tiltrotor Configuration

The tiltrotor configuration employs a conventional fixed-wing airframe with pairs of lateral-tandem rotors mounted at the wingtips. These rotors are actuated to tilt between the vertical and horizontal axes of the airframe. Though current tiltrotor aircraft have a single pair of rotors, designs of even more advanced tiltrotors have more than a single pair of rotors. The need for the actuation of the rotors is to obtain the beneficial performance attributes of both rotorcraft (VTOL, hover, manoeuvrability and omni-directional flight) as well as fixed-wing aircraft (speed, range and altitude) in a single flight vehicle. The primary role of the rotors in *helicopter-type* mode is to provide all the aircraft lift, propulsive and control forces. On the other hand, the primary role of the rotors in *aeroplane-type* mode is to provide solely the propulsive force with the aircraft lift generated by the wing and control forces by the empennage. The amalgamation of a rotorcraft and fixed-wing aircraft into a single flight vehicle expands the flight envelope relative to its individual counterparts and creates a versatile aircraft capable of fulfilling a diverse range of missions with complex requirements, e.g., V/STOL, hover, long range, high speed. The design of tiltrotor aircraft is complex and a large technical effort was required to develop the tiltrotor technology into what we see today. For a thorough history of tiltrotor aircraft covering the technological development of the XV-15 TRRA (TiltRotor Research Aircraft), the reader is referred to [1].

5.1.1   Flight Mode Classification

The unique aspect of tiltrotor aircraft is their ability to operate as either a *rotorcraft* or *turboprop fixed-wing*. We can define three regimes of tiltrotor flight:

1. Helicopter mode
2. Conversion mode
3. Aeroplane mode

These flight modes are somewhat ill-defined and there remains no universal definition between the modes. Classifications could be made based on several factors such as the rotor tilt angle, airspeed, flight characteristics or the share of lift between the rotors and wing. Ultimately, however, *helicopter mode* is synonymous with low-speed flight and *aeroplane-mode* with high-speed flight. Once the aircraft has transitioned to these higher flight speeds, the rotors operate as propellers and the wing lift sustains the aircraft weight. The rotor speed is then reduced to minimise profile power and compressibility effects towards the blade tips (reduced Mach number) and the flap/flaperons are fully retracted to reduce the required power.

5.1.2 The Conversion Corridor

The conversion corridor of tiltrotor aircraft represents a map of possible operating points through the aircraft's transition between helicopter mode and aeroplane mode. It essentially represents the flight envelope of the aircraft presented through the rotor tilt angle τ and airspeed parameters V_∞. The conversion corridor depends on a multitude of factors related to both the aircraft configuration and operating condition. For a given aircraft configuration (weight, flap/flaperon deflection and cg location), the conversion corridor is a three-dimensional surface covering airspeed, rotor tilt and altitude.

The conversion corridor is determined from a set of constraints that limit the validity of the trim state: attitude angle limits; installed power/torque limits; control input limits; structural limits; wing stall. Key parameters that affect the conversion corridor are the weight, required power and altitude. The aircraft weight has the largest effect on the corridor width, increasing both the wing stall speed and the required rotor power and thus, narrowing the corridor width from both sides. Increasing the altitude reduces the installed power from the turboshaft engine and, therefore, further reduces the power-limited airspeed. The installed power at altitude can be modelled from a simple first-order approximation as:

$$P = P_{h=0}\sigma^{0.80} \tag{5.1}$$

where P is the installed power at altitude, $P_{h=0}$ is the installed power at sea level and σ is air density relative to sea level. The flap/flaperon deflections of the wing, discussed more in Sect. 5.2.2, have a significant impact on the required power and have large deflections at low airspeeds for improved performance. These must be retracted to a clean wing configuration for efficient flight at high airspeeds.

The conversion corridor should, ideally, be as wide as possible to maximise the operating space. This would allow the pilot to operate the aircraft at any commanded rotor tilt angle over a wide range of airspeeds. In reality, however, the conversion corridor has a finite range of permissible airspeeds at each rotor tilt angle. If this airspeed range is too narrow, flight at the commanded rotor tilt angle may be challenging and conversion may be proceeded through the flight control system. Therefore, transitional flight must be considered during the design process to ensure the aircraft can safely convert between flight modes with the corridor width maximised.

5.2 Tiltrotor Design Challenges

5.2.1 Hybrid Rotor/Propeller Design

The rotor system of tiltrotor aircraft operates through a diverse aerodynamic environment. This includes a wide range of airspeeds and incidence angles that comprise both axial and non-axial flight conditions. The rotors must be designed to efficiently operate through these diverse operating environments. The distribution of lift along a blade span is a function of the radial location of the blade section r and the angle-of-attack. The section angle-of-attack $\alpha(r)$ is the geometric difference between its pitch angle $\theta(r)$ and inflow angle ϕ:

$$\alpha(r) = \theta(r) - \phi(r) \tag{5.2}$$

The inflow angle is a function of both the kinematic motion of the blade and the induced velocity from the rotor blades and wake. The kinematic motion of the blade is made up of several contributions ranging from the freestream, body rotation rates, rotor tilt rates, rotor speed, flapping and feathering motions. The induced velocity is calculated from an appropriate wake model such as momentum theory or prescribed/free vortex filament models, to name a few.

From Eq. 5.2, we see the twist distribution $\theta(r)$ can be designed and/or augmented to give the desired angle-of-attack distribution and, hence, the spanwise lift distribution. Therefore, given an inflow angle distribution $\phi(r)$ the required twist can be established. However, this relationship is more complex due to the implicit relationship between the rotor loading (controlled by the blade lift and pitch angle) and the wake model. The simplest wake model of a rotor uses momentum theory that relates the steady induced inflow through the rotor λ_i to its thrust C_T:

$$\lambda_i = \frac{C_T}{2\sqrt{\mu^2 + \lambda^2}} \tag{5.3}$$

where μ and λ are the advance and inflow ratios defined as:

$$\mu = \frac{V_\infty}{\Omega R} \sin \alpha_R \tag{5.4a}$$

$$\lambda = \frac{V_\infty}{\Omega R} \cos \alpha_R + \lambda_i \tag{5.4b}$$

where ΩR is the tip speed. The rotor angle-of-attack α_R is defined in terms of the fuselage angle-of-attack α_F and the rotor tilt angle τ:

$$\alpha_R = \alpha_F + (\pi/2 - \tau) \tag{5.5}$$

We may write the velocity of a blade section for an articulated rotor more generally as:

$$\phi(\eta, \psi) = \mathrm{atan} \left(\frac{\lambda + \mu\beta \cos \psi + (\eta - e)\frac{\mathrm{d}\beta}{\mathrm{d}\psi}}{\eta + \mu \sin \psi} \right) \tag{5.6}$$

where β is the blade flapping angle; $\mathrm{d}\beta/\mathrm{d}\psi$ is the blade flapping rate; $\eta = r/R$ is the dimensionless radial ordinate; e is the dimensionless flapping hinge offset, and ψ is the azimuth angle of a blade. In general, we therefore see that the inflow angle

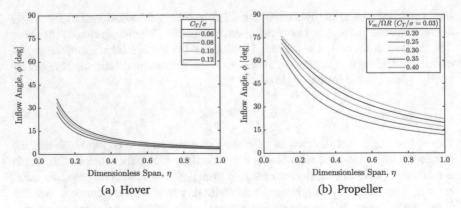

Fig. 5.1 Radial distribution of the inflow angle in hover (at different blade loadings) and in high-speed propeller mode (at constant blade loading)

varies with both the radial position η, azimuth angle ψ and flapping states, β and $d\beta/d\psi$.

In order to simplify Eq. 5.6 and highlight the challenges in tiltrotor blade design, we model the rotor operating in axial flight to remove the azimuthal dependency of the expression:

$$\phi(\eta) = \operatorname{atan}\left(\frac{\lambda}{\eta}\right) \tag{5.7}$$

Figure 5.1 shows the predicted radial inflow angle distribution in steady axial flight for the XV-15 tiltrotor blade at different airspeed ratios and thrust settings. Figure 5.1 highlights the significant differences in the inflow angles between the rotor operating in hover and cruise as a propeller. We see that in order to operate effectively in cruise, a large degree of blade twist is required (and follow a similar distribution to the inflow angle to give the desired angle-of-attack). Comparatively, the twist required in hover is much less and can be well-approximated as a linear twist distribution towards the outboard span stations. This does not adequately approximate the root aerodynamics, however, the inflow angle is not accurately predicted here by simplified momentum theory. Furthermore, the forces and moments produced by the root sections are small compared to the tip region where the dynamic pressure is large and so more inaccuracy is somewhat acceptable in this region. In order to operate effectively at different flight conditions, we see the blade pitch mechanism must be capable of feathering the blades through approximately $\Delta\theta \approx 20°$.

In non-axial flight, the rotor blades experience cyclic loading from the velocity variation around the azimuth (dominated by the $\mu \sin \psi$ term in Eq. 5.6) and any blade dynamics. This generates both a radial and time dependency of the inflow angle as shown by Fig. 5.2. The actual design process of the tiltrotor blades involves a large multi-objective optimisation process with progressively higher-fidelity simulation tools. Typical blade designs have a large degree of twist (that required for propeller

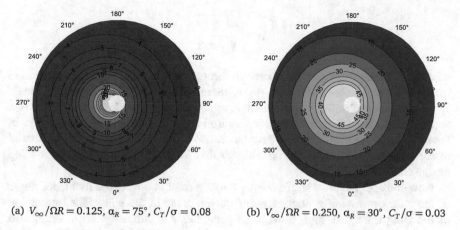

(a) $V_\infty/\Omega R = 0.125$, $\alpha_R = 75°$, $C_T/\sigma = 0.08$ (b) $V_\infty/\Omega R = 0.250$, $\alpha_R = 30°$, $C_T/\sigma = 0.03$

Fig. 5.2 Distribution of the inflow angle around the rotor disc in non-axial flight conditions with the first-harmonic flapping trimmed to zero

mode) with a thick blade root section for improved aerodynamic performance and thinner blade sections towards the tips of the blade where compressibility effects are important.

5.2.2 Rotor/Wing Design

The design of the rotors and wing components of tiltrotor aircraft is an inherently coupled design process. The wing area is sized in order to provide the lift in cruise. A large wing aspect ratio is beneficial for reducing the induced drag, however, poses structural design challenges for tiltrotor aircraft. The fact that the rotors and nacelles are mounted at the tips of the wing ultimately requires a stiff wing structure and, therefore, lower aspect ratio wings are favourable. The shorter wing helps reduce the bending moment at the wing root and also the length of the cross-shafting between the rotors required for engine-inoperative flight.

To avoid the rotor discs overlapping the radii of the rotors are restricted to the length of the wing semi-span. To reduce interactions between the tip vortices and rotor blades and to increase the clearance between the rotors and the fuselage, the rotor radii are reduced relative to the wing semi-span. This design constraint ultimately leads to the rotor thrust being produced over a relatively smaller area compared to conventional helicopter rotors and substantially increases the disc loading T/A ($\equiv$ $W/2A$ for twin lateral-tandem rotors in hover). The disc loading of tiltrotor aircraft can be around $5\times$ larger than lightly loaded utility helicopters and $1.5\times$ larger than heavy-lift helicopters. The downwash velocity induced by the rotor thrust in hover v_i can be related to the disc loading of the rotor:

$$v_i = \sqrt{\frac{1}{2\rho}\frac{T}{A}} \tag{5.8}$$

where ρ is the air density. The downwash through the rotor is, therefore, larger than for conventional rotors and scales proportionally to $\sqrt{T/A}$. This ultimately leads to a large induced power requirement $P_i = v_i T$. The induced power represents a significant portion of the total power required by both helicopters and tiltrotors in hover. Therefore, tiltrotor aircraft have larger power requirements in hover compared to conventional rotorcraft due to the higher disc loadings caused by the rotor-wing design challenges.

Now consider the presence of a wing under a rotor in hover. The incidence angle of the rotor downwash is nominally 90° to the plane of the wing. The immersion of the wing inside the rotor downwash generates a significant vertical drag, also termed a *download*, which must be overcome by additional thrust. This download is detrimental to the payload that can be lifted vertically since additional thrust is wasted overcoming the rotor-induced download. A significant portion of the wing is immersed in the rotor downwash and, due to the large incidence angle, generates download forces that are of the order of 10% of the rotor thrust. We have also shown that rotors with higher disc loadings generate larger downwash velocities. This results in higher dynamic pressure in the wake which ultimately increases the download force. A substantial volume of numerical and experimental research has been conducted to better understand the complex flow physics of the downwash interaction with the wing and help mitigate the download penalty. See [2] for a review of the research into tiltrotor download and the many factors that affect it. The resultant flowfield shows a large region of separation on the pressure side of the wing and requires high-fidelity simulation tools to capture these complex three-dimensional effects [3–5].

Experimental data showed an effective way to reduce the download was to simply minimise the area of the wing projected under the rotor discs [6]. This can be achieved though large deflections of the flap/flaperons (flaps with the dual purpose of augmenting the wing lift and providing roll control) in hover and low-speed flight when the rotors and wing interaction is strongest. Changes to the conventional tiltrotor design have also emerged that negate the download penalty by splitting the wing into three portions: the first portion towards the wing root is fixed relative to the airframe; the remaining portions of the wing (two outboard portions) are fixed relative to the rotor and tilt with it. This design, known as tiltwing, represents a modified tiltrotor configuration. The download is alleviated since the outboard portion of the wing is now parallel, rather than perpendicular, to the rotor downwashes. See Ref. [7] for a brief comparison of tiltwings to tiltrotors.

Consider the rotor shafts in a vertical orientation relative to the airframe. At this orientation, the rotors operate predominately in edgewise flight ($\pm$ trim attitude angles) and, as the airspeed is increased, the rotor wakes become increasingly skewed relative to the rotor shaft axes. Experimental data has shown that as the airspeed is increased, the rotor wakes are swept aft of the wing and by approximately 40 kn there is negligible interaction between the rotor wakes and wing [8]. However, as the rotors are tilted forwards relative to the airframe, the skew angle of the rotors tends to decrease and the wing becomes reimmersed in the rotor wakes. The incidence of the rotor downwashes is reduced compared to that in hover and the magnitude of the downwash decreases rapidly with forward speed. The ramifications of the interaction of the rotor wakes with the wing at these conditions is, therefore, much less. In fact, the interaction of the rotors with the wing can be made favourable by properly selecting the direction of rotation of the rotors to improve the lift-to-drag ratio of the wing [9]. This is a result of the forward tilt of the local lift vector along the wing axis caused by the upwash of the swirl velocities in the rotor wakes.

5.2.3 Whirl Flutter

Whirl flutter is an aeroelastic instability caused by the coupling of the elastic modes of the wing (torsion and bending in both the vertical and chordwise axes) with the excitation of the rotor. Whirl flutter is currently a high-speed limitation of tiltrotor aircraft. Design features used to delay the onset of whirl flutter largely include stiffening the wing and implementing thicker aerofoil sections.

5.2.4 Interaction Aerodynamics

The integration of different aircraft components into a single flight vehicle inevitably leads to interactional effects. These interactions are often implicit, however, first-order interaction effects can often be well approximated by explicit expressions. Furthermore, the severity of the interactions between components depends largely on the operating condition and aircraft configuration, e.g., airspeed, rotor thrust, flap/flaperon setting. The interactional effects between components modifies the aerodynamic forces and moments acting on the aircraft and this can result in changes to several aspects of flight behaviour such as trim parameters, stability of the aircraft about the trim point, aircraft performance, etc. Interaction effects for tiltrotor aircraft are present at a wide range of rotor tilts and airspeeds.

We have so far discussed one significant interaction for tiltrotor aircraft, specifically the rotors-on-wing interaction in hover and low-speed forward flight when the rotors are orientated towards the vertical relative to the airframe. If we consider these rotor tilt angles further, as the airspeed increases the rotor wakes are swept downstream towards the empennage. The pattern of the rotor wake during these non-axial flight conditions forms interlocking helical vortices with the tip vortices rolled-up into strong, dominant vortex structures in the flowfield. The rotors of the tiltrotor aircraft are arranged in a lateral-tandem configuration. In forward flight with no sideslip, the lateral location of the rolled-up tip vortex structure lies approximately $\pm(b - R)$ from the vehicle centreline, where b is the semi-span of the wing and R is the rotor radius. At certain operating conditions (airspeed, rotor angle-of-attack) the rolled-up tip vortex of the rotor lies very close to the empennage and significantly changes the flowfield at the empennage. Early experiments of the XV-15 tiltrotor aircraft demonstrated that at low-speed operating conditions, the rotor wake induced a strong upwash at the tailplane with dynamic pressures nearly twice that of the freestream [8, 10]. Such an interaction led to a reversal of the pilot fore/aft stick gradient with respect to airspeed. Towards higher airspeeds, the rotor wakes were then found to induce a net downwash on the tailplane. These interactions can be modified by altering either the geometry of the empennage surface or its position relative to the rotors.

A lifting wing can be represented as a bound vortex with a circulation strength that varies along its span. The strength of the bound vortex can be related to the lift produced through the Kutta-Joukowski theorem. As the rotors are tilted towards aeroplane mode, the lift production of the aircraft is progressively offloaded from the rotors to the wing. As the lift increases, so does the circulation strength about the bound vortex and a flowfield is developed with an upwash region in-front of the wing and a downwash region aft of the wing. As the rotors are tilted towards aeroplane mode and the wing lift becomes non-negligible they enter its upwash region and, simultaneously, the wing operates within the velocity field induced by rotor wakes [9]. Cumulatively they create time-varying load fluctuations and reciprocate unsteady forces and moments between the two components. The upwash region generated by the wing essentially changes the rotor angle-of-attack that varies in time due to the unsteady loading between the wing and rotors. The longitudinal in-plane force of the rotor was found to increase by around 50% in the presence of the upwash field compared to an isolated rotor case [9]. This unsteady loading leads to vibratory loads with additional contributions due to flowfield disturbances arising similarly from the fuselage.

The lifting wing also creates a downwash region towards the empennage. From fixed-wing theory, the downwash angle at the tailplane is proportional to the lift coefficient of the wing. This downwash is not negligible due to the magnitude of the lift produced by the wing. The downwash angle at the tailplane modifies the effective angle-of-attack of the tailplane and, therefore, must be accounted for for accurate pitching moment predictions. As mentioned previously, an effective way to minimize the download on the wing due to the rotor downwash is to deflect the flaperons. This both reduces the projected area under the rotor and also increases the lift of the wing. Consequently, the downwash angle at the tailplane is significant, even in low-speed flight with the rotors towards the vertical of the airframe.

At low airspeeds the wing lift is small and the rotors produce the aircraft lift. At these low airspeeds and when the rotors are tilted forwards relative to the airframe, the rotor thrust vectors (parallel to the rotor shaft axes) are also tilted forwards and the component of the lift relative to the inertial vertical is reduced. Therefore, the aircraft pitches nose-up to realign the thrust vector against the weight vector [11]. This behaviour further compounds the flap effects on the downwash and increases the lift of the wing through an increased fuselage angle-of-attack. The downwash generated by the wing increases further and exemplifies the significance of the empennage in determining the aerodynamic moments at low airspeeds.

We have so far discussed only the interaction phenomena that affect the longitudinal motion of the aircraft. Tiltrotor aircraft are also susceptible to lateral/directional interactions when the aircraft enters sideslip conditions. Consider the rotors in hover and low speed with the rotors orientated towards the vertical relative to the airframe. At these conditions, the wing is immersed in the wakes of the rotors and a download is felt on the airframe. When sideslip conditions are introduced, this causes unequal portions of the wing to be immersed in the rotor wakes and this leads to a net rolling moment by the wing. At higher airspeeds, the increased freestream convects the rotor wakes downstream towards the empennage. We have discussed how the rolled-up tip vortices of the rotor led to a strong upwash over the tailplane at certain operating conditions which resulted in undesirable handling qualities (stick position reversal with airspeed). In a similar fashion, the introduction of sideslip (either positive or negative) skews the wake of the windward rotor towards the fuselage centreline. Depending on the position of the empennage relative to the rotor, this skewed rotor wake may sit over the tailplane and induce a strong downwash that leads to a nose-up pitching moment. This phenomenon is known as *pitch-up-with-sideslip* in tiltrotor literature [12]. The exact interaction mechanism is dependent on the airspeed, rotor tilt angle, sideslip angle, blade loading and the position of the empennage relative to the rotor. There are similar effects at the empennage from the lateral displacement of the wing wake, however, these effects are less pronounced compared to the rotor wake effects.

5.3   Control Methodology

Conventional rotorcraft with a single main rotor augment the rotor forces through three control inputs: collective pitch θ_0 applied to all the blades simultaneously; longitudinal cyclic pitch θ_{1s} and lateral cyclic pitch θ_{1c}. The pitch angle of a blade section θ is described by a first-harmonic Fourier series:

$$\theta(r, \psi) = \theta_{tw}(r) + \theta_0 + \theta_{1s} \sin \psi + \theta_{1c} \cos \psi \tag{5.9}$$

where $\theta_{tw}(r)$ is the built-in blade twist distribution as a function of span. The one-per-rev variation in blade pitch generates cyclic forces that flap the rotor blades and tilt the resultant force vector. The magnitude of the force is predominantly augmented by the collective pitch with the cyclic pitch controls predominantly used to provide the directional control of the force vector. On the other hand, conventional fixed-wing aircraft typically use simpler control surface deflections to generate the control forces and moments. Roll control is augmented through asymmetric aileron deflections located at the outboard stations of the wing, pitch control is augmented through elevator deflection on the tailplane and yaw control is augmented through rudder deflection located on the vertical fin. A conventional tailplane and rudder have been

Table 5.1 Aircraft response to differential rotor control inputs at different rotor tilts

Differential control	Rotors vertical	Rotors horizontal
Collective	Roll moment	Yaw moment
Longitudinal cyclic	Yaw moment	Roll moment
Lateral cyclic	Roll moment	Yaw moment

described here, however, pitch and yaw control are provided by some arrangement of aerodynamic surfaces constituting an empennage.

The existence of two control methods, rotary-wing and fixed-wing controls, requires additional considerations when designing the control system. Controls are typically sized and adequate deflections determined to provide sufficient control forces and handling qualities throughout the flight envelope. However, for the tiltrotor configuration there are several strategies to augment the required control axis. Similar to aileron control of fixed-wings, differential control can be implemented between the rotors. First consider differential collective pitch. This acts to increase the thrust on one rotor and decrease that on the other, thereby generating asymmetric forces and a net moment. This differential control input generates a moment about either the roll or yaw axis, or both, depending on the tilt angle of the rotors. Now consider differential cyclic inputs. These differential control inputs act to generate resultant moments through asymmetric flapping responses. The effects of differential rotor controls are summarised in Table 5.1.

The authority of the fixed-wing control surfaces is dependent on the dynamic pressure and, hence, airspeed. As a result, the ability of fixed-wing control surfaces to provide control moments at low speed, including hover, is minimal and rotary-wing controls must be utilised instead. As the airspeed and dynamic pressure increase, the authority of the fixed-wing control surfaces surpasses the rotary-wing control authority and the cyclic controls become mostly redundant. The requirement of both sets of controls at different operating points introduces a scheduling requirement of the fixed-wing and rotary-wing controls. Fundamentally, multiple control states for a single control axis results in an under-determined control system. There is, therefore, no unique solution to the control problem of tiltrotor aircraft and it is up to the control engineer to design a control system that ensures adequate control characteristics are available throughout the flight envelope.

The pilot does not control the individual rotary-wing and fixed-wing controls, rather the position of the stick, pedals and throttle/power/collective lever. Displacements of the pilot controls are transferred to control system inputs to give expected aircraft behaviour, i.e., pushing the stick forwards gives a nose-down tendency and right pedal gives a yaw to starboard. We focus now on control about the pitch axis. The control moments about this axis are generated through longitudinal cyclic applied to the rotors and elevator deflection applied to the tailplane. A generic control system for tiltrotor aircraft relating these control inputs to the pilot fore/aft stick can be expressed as:

$$\theta_{1s} = A_{\theta_{1s}} \delta \frac{\partial \theta_{1s}}{\partial \delta} \tag{5.10a}$$

$$\eta = A_\eta \delta \frac{\partial \eta}{\partial \delta} \tag{5.10b}$$

where $A_{[\,]}$ is the control input amplitude of the subscripted quantity (which can be positive or negative depending on the sign convention of the control input); δ is the fore/aft pilot stick position; and $\partial \theta_{1s}/\partial \delta$ and $\partial \eta/\partial \delta$ are the longitudinal cyclic and elevator control gearings with respect to the stick position. Similar expressions to Eqs. 5.10 can be used for roll and yaw controls. Here, the stick position is a normalised quantity in the interval $-1 \leq \delta \leq 1$ with $\delta = -1$ being full aft, $\delta = 0$ implying a neutral stick position (no pilot control input) and $\delta = 1$ being full fwd. The control gearings have values in the interval $0 \leq \partial[\,]/\partial \delta \leq 1$ where $[\,]$ represents an arbitrary control input variable. A control gearing of $\partial[\,]/\partial \delta = 0$ implies the control input is fully disengaged from the pilot stick position and a gearing of $\partial[\,]/\partial \delta = 1$ implies the control input is fully engaged. The control gearing functions, $\partial \theta_{1s}/\partial \delta$ and $\partial \eta/\partial \delta$, are arbitrary and are determined by the control engineer to ensure good control is available throughout the flight envelope. Typical control gearing functions are linear functions or trigonometric functions based on the tilt angles of the rotors [13]. More complex functions could also be introduced depending on other flight parameters such as airspeed, altitude, etc. Furthermore, these control gearing functions could also be found through an optimisation routine.

5.4 Modelling and Simulation

Predicting the flight behaviour of any aircraft requires adequate engineering models describing their flight mechanics and dynamics in time and three-dimensional space. In this context, we take the term *aeromechanics* to mean the integration of *aerodynamic* models for different aircraft components into a main *flight mechanics* model describing the translational and rotational motions of the complete flight vehicle. This can be further extended to include *aeroelastic* effects from structural behaviour. Aeromechanics models are built on multi-fidelity physical models of aerodynamic flows, dynamics models as well as structural models. The implementation of an aeromechanics model depends on the required fidelity of the problem at hand. For example, initial aircraft design is performed using reduced-order models at little computational cost which can be enriched with experiential engineering (previous experience, experimental or higher-order numerical data). Aeromechanics models can then be established to assess trim behaviour, stability and control, aircraft performance and control system design, to name a few. As the design progresses, higher fidelity tools are implemented to refine and optimise component designs. Aeromechanics models are, therefore, wide-reaching within the aerospace discipline right through the life cycle of an aircraft from conceptual design to pilot training on flight simulators.

5.4.1 Generic Flight Mechanics Model

In order to understand the flight mechanics of tiltrotor aircraft we first establish a flight mechanics model. In keeping with the literature of both flight mechanics and flight dynamics, we model the aircraft as a point-mass rigid-body and attach the conventional body-fixed Cartesian system to the centre of gravity. Since the coordinate system is fixed to the body and is free to accelerate in translation and rotation it is, therefore, a non-inertial frame of reference. The equations of motion are derived from Newton's law of motion assuming a constant weight:

$$\vec{F} = m \frac{\mathrm{d}\vec{v}}{\mathrm{d}t} \tag{5.11a}$$

$$\vec{M} = \frac{\mathrm{d}\vec{h}}{\mathrm{d}t} \tag{5.11b}$$

where $\vec{F}$ and $\vec{M}$ are the external forces and moments; m is the aircraft mass (assumed constant); $\vec{v}$ is the linear velocity; $\vec{h}$ is the angular momentum; and $\mathrm{d}/\mathrm{d}t$ is the temporal derivative. The equations of motion are typically derived based on a rigid-body approximation of the aircraft, i.e., no elastic deformation of the components and/or relative motion between components. The former is a valid approximation for tiltrotor aircraft on the assumption that all components are rigid or that elastic deformations are small enough to be negligible (both the rotor blades and wing are very stiff in tiltrotor designs required for high-speed flight). The latter is not strictly true for two reasons:

1. The rotors have rotational inertia themselves, however, are counter-rotating and therefore cancel each other out in steady longitudinal flight conditions.
2. The rotors are able to tilt in-flight through pilot command to transition between low-speed and high-speed flight. This dynamic tilting is not a negligible effect due to the large masses of the rotors and nacelles, however, the assumption of static components (rotors fixed relative to the airframe) simplifies the analysis.

The external loads $\vec{F}$ and $\vec{M}$ comprise of the *aerodynamic* loads generated by the different aircraft components as well as the gravitational loads. The challenge is now to determine these aerodynamic loads. The conventional tiltrotor aircraft consists of two lateral-tandem rotors mounted onto the airframe at the wingtips. We may then decompose the aerodynamics loads in the airframe loads, comprising contributions from the fuselage, wing and empennage, and the rotor loads. A generic load L may then be written as the following sum:

$$L = L_F + L_W + L_E + L_R \tag{5.12}$$

where the subscripts F, W, E and R denote the fuselage, wing, empennage and rotors. In addition to the discretisation detailed in Eq. 5.12, the nacelles that house the turboshaft engines for the rotors are significant in terms of both their size and mass. Consequently, the aerodynamic forces associated with these nacelles, particularly the drag, are not negligible at many operating points with a large area projected against the freestream. As a result, the load decomposition is rewritten to include the nacelle loads:

$$L = L_F + L_W + L_E + L_R + L_N \tag{5.13}$$

where, similarly, the subscript N denotes the nacelle loads.

We have so far established a basic flight mechanics model of a generic tiltrotor aircraft and decomposed the loads into airframe and rotor contributions. The challenge now is to implement aerodynamic models for each of these load sources. These aerodynamic models depend on the component being analysed, its geometry and aerodynamic environment (which may depend on other components), the control inputs applied and the required fidelity of the model.

5.4.2 Vehicle Geometry

In defining an aeromechanics model, the spatial position of the different aircraft components are required when evaluating the moments acting on the aircraft. When using a body-fixed frame of reference whose axes are located at the cg, the weight produces no moment. However, the cg is not a well-defined point on the aircraft and varies with the operating conditions. Furthermore, the tilting of the rotors and their nacelles causes displacements of the cg along the longitudinal and vertical axes of the aircraft. Given that the weight of the rotors and nacelles is a significant portion of the total weight, the displacement of the cg is not negligible. Dynamic effects of the cg (velocity and acceleration) due to rotor tilt motions should also be included in higher-fidelity flight dynamics models for an improved representation of the vehicle dynamics.

In order to define the layout of the aircraft we implement an alternative frame of reference that is independent of the vehicle's cg location. We define this as the vehicle frame of reference with vehicle axes pointing nose-to-tail (x-axis), port-to-starboard (y-axis) and upwards (z-axis). In this frame of reference, we define the location of all the aircraft components and the cg. Inside the flight mechanics model when the aerodynamic moments are calculated, the position of the aircraft components relative to the cg are computed based on the instantaneous locations. The origin of the vehicle axes is arbitrary, however, is typically located at or in-front of the nose of the aircraft and along the vertical plane of symmetry of the fuselage.

5.4.3 Overview of Component Aerodynamics

Fuselage Aerodynamics

Accurately determining the aerodynamic loads generated by the fuselage are difficult due to the influences of the rotor and airframe interactions. The tilting degree of freedom only acts to make the load prediction even more complex as an additional degree of freedom is introduced. Sideslip conditions may also significantly affect the lateral/directional loads due to the large rotor wakes and their possible impingement on the fuselage. The aerodynamics of the fuselage are particularly important at higher airspeeds when the parasitic power is greatest. The aerodynamic loads generated by the fuselage are usually determined from wind-tunnel experiments, panel methods or high-fidelity simulation tools. In the first instance the design of the fuselage is not typically known. The fuselage aerodynamic drag is a significant factor at high-speed flight and cannot, therefore, be excluded from design calculations. First-order approximations of the fuselage parasitic power are typically made from historical trends of similar aircraft.

Rotor Aerodynamics

Rotor aerodynamics are implicitly complex due to the rotating nature of the blades. Most aerodynamic models that have been developed for helicopter rotors are also applicable to the rotor systems of tiltrotor aircraft. However, there are a couple of significant differences that will be mentioned and these should be addressed in the development of any aerodynamic model for tiltrotors. Firstly, unlike conventional rotors, the incidence angle of the rotors in edgewise flight is not small. In high-speed flight, this ultimately means the inflow ratio through the rotors is large and the simplification of *small inflow angles* (see Sect. 5.2) is not applicable for the analysis of the blade loads. Secondly, wake models developed under the guise of the rotors being lightly loaded should be verified for tiltrotor systems due to the high disc loadings during helicopter mode operation. The requirements of the rotor aerodynamic model should match the problem being investigated.

Wing Aerodynamics

The aerodynamics of the wing are made more complex by the rotors and the interaction of the rotor wakes with the wing and the wing wake. The requirements of the wing aerodynamic model must capture the significant effect of the download in hover and model its transition into forward flight. Furthermore, if the aircraft enters sideslip conditions this may result in unequal areas of the wing being immersed in the rotor wakes. As such, the basic requirement for the wing aerodynamic model is at least the prediction of the spanwise loading through techniques such as lifting-line or lifting-surface.

Empennage Aerodynamics

Modelling the empennage component is the most challenging component due to the significant interactions from both the airframe and rotor components. The exact interaction mechanism is dependent on the flight condition. The rotor wake interaction is dominated by the freestream velocity, the rotor incidence angle and the blade loading. The wing interaction is dominated by the lift coefficient of the wing and is, therefore, dependent on any flap/flaperon deflections. Sideslip effects on the empennage generate sidewashes that can also couple into the longitudinal degree of freedom (pitch-up-with-sideslip [12]). All interactions of course depend on the relative position between the components. The role of the empennage is to produce control moments about the pitch and yaw axes and, therefore, accurate load predicts can significantly affect both the predicted steady and unsteady performance of the aircraft.

Nacelle Aerodynamics

The geometry of the nacelles is dictated by that of the turboshaft engine it houses. This can result in an nonaerodynamically shaped body that produces a significant amount of drag. For tiltrotor aircraft, this eventually is realised when a large frontal area of the nacelle geometry is project against the freestream, i.e., forward flight with the rotor nacelles tilted towards the vertical. Consequently, an increase in the rotor thrusts are required to overcome the drag of the nacelles and this is accompanied by an increase in required power. The nacelle drag can be modelled using a simple *equivalent flat-plate* type model, however, this should at least include a dependency on the incidence angle of the nacelle to account for its tilt through the flight envelope.

5.4.4 Pilot Gearings

The next part of building a simulation model are the pilot gearings, specifically, how the pilot control inputs are related to physical control inputs of the rotors and airframe. As discussed in Sect. 5.3, this system must be designed/optimised for adequate control and handling through the flight envelope. This is not a trivial task, though simple gearing functions have been proposed in literature [13].

5.4.5 Key Operating Parameters

We now briefly discuss the key operating parameters that affect performance, particularly during transition. The aerodynamic environment is strongly coupled to the airspeed, rotor tilt angle and flight path angle. The required power scales proportionally to the cube of the airspeed and, thus, high-speed flight is limited by the installed power of the aircraft. Climbing also increases the required power, though causes the induced power to drop due to the increased inflow through the rotor disc. Altitude effects are significant due to both a reduced density and net reduction in installed power.

In terms of the aircraft configuration, by far the most important parameter is weight. This both increases the stall speed of the wing and increases the required power for the rotors. Both these influences have a *pincer-like* effect on the conversion corridor boundaries, narrowing the operational envelope. Flap/flaperons spanning the entire wing length are used to minimise the rotor-induced download on the airframe, increase the $C_{L_{max}}$ of the wing (thus decreasing the stall speed) and provide roll control. The increase in lift is not for free and there is a considerable increase in the lift-induced drag and, therefore, required power. Unlike the weight effect, the flap/flaperon effects on the conversion corridor are to push the envelope to lower airspeeds as opposed to a pincer effect. The final key parameter is the centre of gravity location. Every aircraft has a permitted centre of gravity range in which the aircraft is controllable. This is the same for tiltrotor aircraft, however, the displacement of centre of gravity with respect to the rotor tilt must also be considered.

5.5 Numerical Example

We now consider a numerical example of tiltrotor aeromechanics through the conversion corridor and look at its performance and trim characteristics. The aircraft simulated is the Bell XV-15 that has been widely documented in literature. TARA (Tiltrotor AeRomechanics Analysis), an aeromechanics code developed through the UK Vertical Lift Network, was used to run the simulations. The parameters of the simulation model were largely derived from the GTRS model of the XV-15 [14]. The flight conditions used in the present simulations were for steady level flight conditions at ISA sea level. These flight conditions are summarised in Table 5.2 and consist solely of longitudinal conditions. The numerical predictions are compared to those of the GTRS model to validate the simulation model [15]. The constraints imposed on the trim solution are summarised in Table 5.3.

Table 5.2 Simulated flight conditions

Flight path angle	$0°$
Altitude	Sea level
Weight	5900 kg
CG location	Aft limit
Rotor speed	589 rpm
Flap/flaperon deflections	$40.0°/25.0°$ ($\tau \leq 15°$)
	$25.0°/12.5°$ ($\tau > 15°$)

Table 5.3 Performance constraints imposed on the trim solution

Fuselage pitch attitude	$|\theta_F| \leq 15°$
Fore/aft stick position	$|\delta| \leq 1$
Fore/aft flapping	$|\beta_{1c}| \leq 12°$
Installed power	$C_P/\sigma = 0.0144$ (single rotor)

Fig. 5.3 Predicted
conversion corridor of the
XV-15 tiltrotor against
literature data

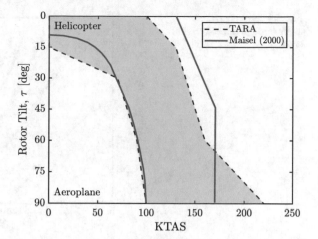

The conversion corridor predicted by TARA is shown in Fig. 5.3 and is compared to the literature corridor [1]. Good correlation is observed between the two corridors overall. TARA under-predicts the high-speed boundary compared to the published corridor up to a rotor tilt angle of 60° and over-predicts it thereafter. Firstly, this discrepancy is accountable due to the nonspecific configuration parameters used to define the reference corridor and secondly, the omittance of a structural constraint within TARA. The latter is what imposes the conversion high-speed limit of the XV-15 at around 170 kn [16] and highlights the need for future structural modelling. Better agreement of the conversion corridor would likely occur if the flap/flaperons were retracted towards the high-speed boundary, delaying the power constraint to higher airspeeds.

The pitch of the aircraft through transition is shown in Fig. 5.4. The tendency of the aircraft pitch with airspeed is to pitch nose-down. When the rotor are towards vertical, the nose-down tendency provides the propulsive force to overcome the parasitic drag of the airframe. This is the same behaviour as classical rotorcraft. The nose-down tendency is proportional to the drag and, therefore, both the drag of the nacelles and deployment of high-lift devices exacerbates this behaviour. Increasing the rotor thrust has the opposite effect, reducing the nose-down tendency with airspeed. Excessive nose-down pitch attitudes may be encountered at higher airspeeds in this helicopter-type mode of flight. When the aircraft lift is generated largely from the wing, a nose-down tendency with airspeed is still observed. This is caused by a reduction in the wing C_L since the dynamic pressure is larger. This is classical fixed-wing behaviour. Contrary to helicopter-type mode of flight, excessive positive pitch angles may be encountered at lower airspeed where the largest wing C_L is required.

The effect of the rotor tilt on the pitch angle at low airspeeds is reactive [11]: as the rotors tilt forwards, the fuselage pitches nose-up against the rotor tilt, as shown by the curves for $\tau = 0°$ and $\tau = 15°$ in Fig. 5.4. This behaviour is observed to tend the rotor thrusts against the weight vector and ensure vertical force equilibrium

Fig. 5.4 Fuselage pitch
angle through the conversion
corridor

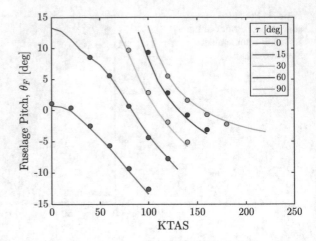

is achieved. There is, therefore, a coupling between these degrees of freedom at low airspeeds. The orientation of the rotor thrust vector can be augmented not only through the rotor tilt, but also through cyclic control that flaps the rotor. Therefore, we also have a coupling of the fuselage pitch with the rotor tilt angle and rotor flapping. Consequently, the rotor flapping is a function of the cyclic pitch inputs which are related to the pilot stick position through some gearing mechanism. In fact, from a horizontal force balance the fuselage pitch angle can be shown to be given approximately by

$$\theta_F = \tan \tau_{TPP} - \frac{D}{nW \cos \tau_{TPP}} \tag{5.14}$$

where $\tau_{TPP} = \tau + \beta_{1c}$ is the tip-path plane tilt relative to the body vertical; D is the airframe drag; $n = T/W$ is the thrust-to-weight ratio; and W is the aircraft weight. The airframe drag contains both the parasitic contribution and the lift-induced contribution and is, therefore, largely dependent on the fuselage angle-of-attack and wing flap/flaperon deflections.

The longitudinal rotor flapping through conversion is shown in Fig. 5.5 and shows only a moderate amount of flapping was required throughout the corridor. The largest amount of flapping was encountered in hover with the rotors tilted forwards by 15° due to the necessity to orientate the thrust vector against the weight vector. In hovering flight for $\tau = 0°$, a small forward tilt of the tip-path plane was required to trim out the pitching moment from the weight and wing download about the rotor pivot point at the aft cg position. The required flapping in hover must trim out the pitching moments from the aircraft weight and, also, the rotor-induced download over the wing. The correlation of the flapping with the GTRS model matches favourably. The rotor flapping is significantly affected by the wake interactions of the rotors and wing with the empennage at low-speed due to trim control being generated by flapping.

The fore/aft stick position through conversion is shown in Fig. 5.6. The stick position shows a transition from typical rotary-wing to fixed-wing behaviour with

Fig. 5.5 Fore/aft rotor flapping through the conversion corridor

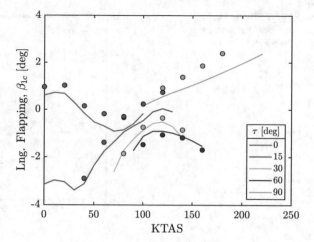

Fig. 5.6 Fore/aft pilot stick position through the conversion corridor

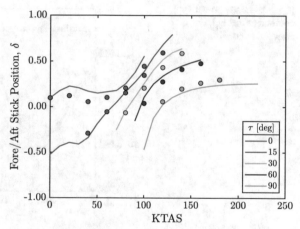

respect to the rotor tilt angle. At around 40 kn a reversal of the stick position gradient with airspeed is observed and results from the upwash induced by the rotor wakes at the tailplane. The flight conditions and aircraft geometry play an important role to determine which operating points this interaction is most pronounced. The stick position is implicitly linked to the control gearings to determine the control inputs to both the fixed-wing and rotary-wing components.

The blade loading of a single rotor through transition is shown in Fig. 5.7. The blade loading demonstrates there is an order of magnitude difference in the thrust requirements between helicopter and aeroplane modes. There are also variations in blade loading with airspeed. This is due to the multifunctional requirement of the rotors to provide the lift, propulsive and control forces and the variation in aircraft drag. Towards aeroplane mode, the blade loading scales with the airframe drag since the rotors operate as almost pure thrust producers: lower airspeeds have larger lift-induced drag due to higher C_L requirements from the wing; higher airspeeds have

Fig. 5.7 Blade loading
coefficient through the
conversion corridor

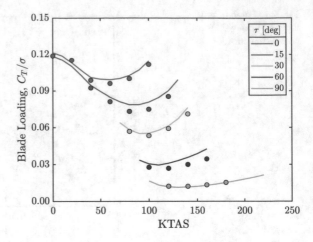

Fig. 5.8 Blade power
coefficient through the
conversion corridor

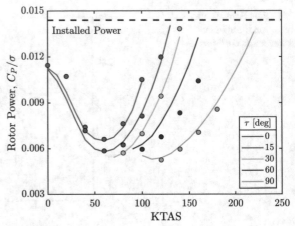

lower lift-induced drag but higher parasitic drag. In fact, this behaviour is observed
for all rotor tilt angles beyond roughly 80 kn.

The variation in blade power coefficient is shown in Fig. 5.8. The power curves follow the typical behaviour of rotorcraft: large power requirements in hover due to the
induced power (amplified for tiltrotors due to the much greater disc loading); reduced
power thereafter with increasing airspeed due to the reduction in induced power; followed by an increased in power due to the parasitic drag of the airframe at higher
airspeeds (scaling with V_∞^3). The effect of the rotor tilt is to increase the airspeed
for minimum power. As seen from Fig. 5.8, the required power is a constraining performance factor at high-speed flight. Therefore, the flap/flaperons need be retracted
in order to help minimise the power requirements and postpone the power-limited
flight to higher airspeeds. Structural considerations also become important at higher
airspeeds.

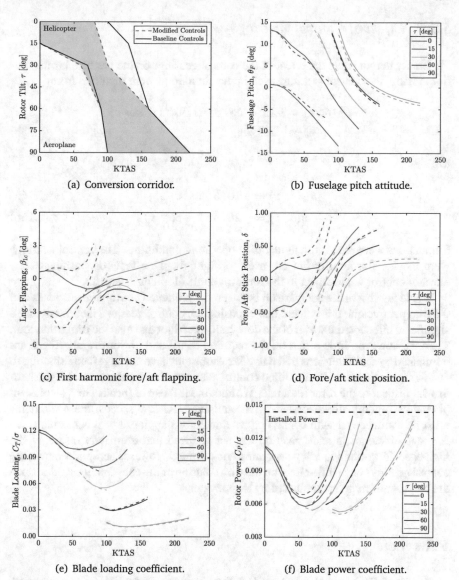

(a) Conversion corridor.

(b) Fuselage pitch attitude.

(c) First harmonic fore/aft flapping.

(d) Fore/aft stick position.

(e) Blade loading coefficient.

(f) Blade power coefficient.

Fig. 5.9 Modified control gearing simulations

5.5.1 Effect of Control Gearings

We briefly demonstrate the effect of the control gearings on the predicted trim solution. Firstly, the control gearings used in the simulation were modified from:

$$\theta_{1s}(\delta, \tau) = -10° \, \delta \, \cos \tau - 1.50°(1 - \cos \tau) \tag{5.15a}$$

$$\eta(\delta) = \quad 20° \, \delta \tag{5.15b}$$

to

$$\theta_{1s}(\delta, \tau) = -10° \, \delta \, \cos^2 \tau \tag{5.16a}$$

$$\eta(\delta, \tau) = \quad 20° \, \delta \, (1 - \cos^2 \tau) \tag{5.16b}$$

Simulations were then rerun at identical operating conditions. The control gearings were modified arbitrarily but are used to highlight the effects of washing-in the elevator control with respect to the rotor tilt angle ($1 - \cos^2 \tau$ term).

Figure 5.9 shows the comparison between the predicted conversion corridors and the trim parameters. It is evident that choosing the control gearings incorrectly has a significant effect on the width of the conversion corridor and also the trim behaviour. There is, however, little effect on the required thrust and power overall. These are dominated by the deflections of the flap/flaperons that have the largest contribution to an increase in drag. For the modified control gearings, the elevator control is washed-in with respect to the rotor tilt angle. Without the full use of the elevator, significant pitching moments are generated by the tailplane. At low speeds, these moments must be trimmed out with cyclic control and rotor flapping. For this reason, there is a steep increase in stick position against airspeed that ultimately results in the high-speed corridor boundary becoming stick limited. Towards aeroplane mode, the control gearings have little effect on the trim solution or the conversion corridor since the control authority is dominated by the elevator.

References

1. Giulianetti DJ, Maisel MD, Dugan DC (2000) The history of the XV-15 tilt rotor research aircraft: from concept to flight. SP-2000-4517. NASA
2. Felker F (1988) A review of tilt rotor download research. In: 14th European rotorcraft forum, Milan
3. Lim J (2019) Fundamental investigation of proprotor and wing interactions in tiltrotor aircraft. In: Vertical flight society 75th annual forum & technology display, Philadelphia
4. Tran S, Lim J, Nunez G, Wissink A, Bowen-Davies G (2019) CFD calculations of the XV-15 tiltrotor during transition. In: Vertical flight society international 75th annual forum & technology display, Philadelphia
5. Tran S, Lim J (2020) Investigation of the interactional aerodynamics of the XV-15 tiltrotor aircraft. In: Vertical flight society international 76th annual forum & technology display

6. McCroskey W, Spalart P, Laub G, Maisel M, Maskew B (1983) Airloads on bluff bodies, with application to the rotor-induced downloads on tilt-rotor aircraft. NASA TM 84401
7. Nannoni F, Giancamilli G, Cicala M (2001) ERICA: the European advanced tiltrotor. In: 27th European rotorcraft forum, Moscow
8. Wilson J, Mineck R, Freeman C (1976) Aerodynamic characteristics of a powered tilt-proprotor wind-tunnel model. NASA TM X-72818
9. McVeigh M, Grauer W, Paisley D (1990) Rotor/airframe interactions on tiltrotor aircraft. J Am Helicopter Soc 35(3):43–51
10. Marr R, Sambell K, Neal G (1973) V/STOL tilt rotor study. In: Volume 6: hover, low speed and conversion tests of a tilt rotor aeroelastic model. NASA CR 114615
11. Appleton W, Filippone A, Bojdo N (2021) Interaction effects on the conversion corridor of tiltrotor aircraft. Aeronautical J 1–22
12. He C, Zhao J (2012) High fidelity simulation of tiltrotor aerodynamic interference. In: American helicopter society 68th annual forum
13. Dreier M (2007) Introduction to helicopter and tiltrotor flight simulation, AIAA
14. Ferguson S (1988) A mathematical model for real time flight simulation of a generic tilt-rotor aircraft. NASA CR 166536
15. Ferguson S (1989) Development and validation of a simulation for a generic tilt-rotor aircraft. NASA CR 166537
16. Maisel M (1975) Tilt rotor research aircraft familiarization document. NASA TM X-62 407

Wesley Appleton having pursued a Ph.D. at the University of Manchester on tiltrotor aeromechanics, is now working at the Defense Science and Technology Laboratory (DSTL) of the UK Ministry of Defense, in Fareham, Hampshire.

Chapter 6
Rotor Acoustics

Dale Smith and George Barakos

Abstract This chapter explores selected aspects of rotor acoustics. With increasingly stringent certification requirements, this field is becoming steadily more important in the design of rotorcraft. In this chapter, we discuss the importance of studying rotorcraft acoustics and the process used by regulators to evaluate their noise emissions. We then introduce the various noise sources and their importance in relation to the vehicle's total noise emissions. We proceed by introducing Lighthill's acoustic analogy and subsequently the Ffowcs Williams and Hawkings equation. Based on these, we describe a number of approaches to estimate the contributions of the various noise sources. We describe the implementation of an acoustic code and accompany with sample MATLAB scripts to evaluate the thickness and loading sources of an arbitrary rotor. We conclude the chapter by looking at a range of methods to reduce rotorcraft noise sources and discuss the future outlook for rotorcraft acoustics.

Nomenculture

$\Box$	Wave operator
A_p	Panel area, m^2
B	Rotor blade count
BPF	Blade passage frequency $= nB$, Hz
c_0	Speed of sound, m/s
EPNL	Effective perceived noise level, EPNLdB
f	Surface definition
H	Heaviside function
l	Vector of blade forces, N/m

Supplementary Information The online version contains supplementary material available at https://doi.org/10.1007/978-3-031-12437-2_6.

D. Smith
QinetiQ, Rosyth, Scotland

G. Barakos (✉)
Glasgow University, Glasgow, Scotland
e-mail: george.barakos@glasgow.ack.uk

MTOM	Maximum takeoff mass, kg
M	Mach number
n	Rotor rotational frequency, rev/s, or unit normal
p'	Acoustic pressure, Pa
p_S	Panel surface pressure, Pa
PAV	Personal air vehicle
r	Blade sectional radius, m
$\vec{r}$	Radiation vector, m
R	Magnitude of radiation vector, m
$\mathscr{R}$	Source-observer distance, m
SEL (or L_{AE})	Sound exposure level, dBA
SPL	Sound pressure level, dB
t	Observer time, s
T_{ij}	Lighthill's stress tensor, kg/ms^2
u	Velocity component, m/s
V_∞	Free-stream velocity, m/s
UAM	Urban air mobility
$\vec{V}$	Blade velocity components, m/s
$\vec{x}$	Coordinate vector of observer location, m

Greek Symbols

δ	Kronecker delta or Dirac delta function
Θ	Angle between normal and radiation directions, rad
ρ	Fluid density, kg/m^3
τ	Source time, s
τ_{ij}	Viscous stress tensor
Ω	Rotor rotational speed, rad/s

Subscripts

d	Denotes derivative with respect to source time (MATLAB example)
L	Loading component
n	Component in normal direction
r	Component in direction of observer
T	Thickness or blade tip component

Other

[']	Derivative with respect to time
[¯]	Generalised derivative
[⌐]	Variable has vector components

6.1 Introduction

Rotorcraft control is a complex problem, with flapping and lateral and longitudinal cyclic required to balance the forces and moments about the vehicle. As a result, the main rotor blade is faced with a complex aerodynamic environment that constantly changes as the blade rotates. One by-product of this aerodynamic environment is significant acoustic emissions. These acoustic emissions are of particular concern for manufacturers, regulators, operators and the community.

Although significant progress in our understanding and control of rotorcraft noise has been made over the past decades, they remain a serious concern for the industry. In fact, in contrast to the earlier days of rotorcraft development where performance was the sole driver for the design, the acoustic emissions now form a major part of the rotorcraft design. Whilst this is primarily driven by the need to meet regulatory requirements, there are a number of additional pressures which drive the focus on reducing noise.

Whilst rotorcraft must meet regulatory noise requirements prior to their introduction to the commercial market, their commercial success lies firmly in the hands of the public perception and acceptance of the vehicle. Unfortunately, rotorcraft generally have poor acceptance, particularly with regards to their noise emissions. Figure 6.1 shows the results from a CAA study related to public perception of rotorcraft noise. The results from the study paint a poor picture for rotorcraft. The study found that small helicopters were found to have similar levels of annoyance to that of fixed-wing aircraft despite an approximately 15 dB reduced level of noise. The reasons for the increased annoyance lies with the public's perception of helicopters as inherently noise machines.

Therefore, whilst a manufacturer can spend a great deal of resources in reducing the noise level of their rotorcraft, it may be to no end if it is received poorly by the public. As a result, manufacturers must also pay a great deal of attention to the noise perception of their vehicle. This will be of particular concern if we are to see a large uptake in rotorcraft as the Personal Air Vehicle (PAV) market begins to takeoff.

Noise reductions may be also be required as part of the rotorcraft's mission requirements. For example, military rotorcraft may have operational stealth as a design requirement. In which case, the acoustic emissions will be of particular concern for the design team.

Rotorcraft noise is both an environmental and design challenge. Rotorcraft noise is a complicated problem that is described by a number of different sources. These sources cover a range of directivity directions and spans across a wide frequency range. Nonetheless, in order to meet the requirements of reducing noise, we must first form a better understanding of these sources and develop the capability to predict them. In this chapter, we will first look at the requirements set by the regulators on rotorcraft noise. We will then introduce the various rotorcraft sources and discuss their relative importance. Further, we will then look at a number of methods to evaluate these sources. Finally, we will look at a number of approaches to reduce rotorcraft noise.

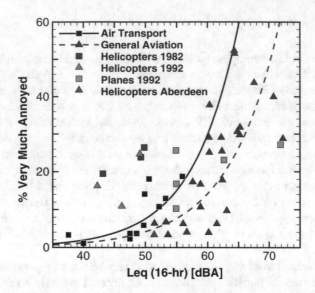

Fig. 6.1 Perceived aircraft annoyance with respect to measured noise, adapted from [1]

6.2 Noise Certification

Prior to entry into service, rotorcraft must first be certified against a range of standards, including their noise emissions. These standards are set by regulatory bodies, for example, the International Civil Aviation Organisation (ICAO), the Federal Aviation Authority (FAA), European Union Aviation Safety Agency (EASA), etc. Rotorcraft noise limits are set by the ICAO, from which most national regulatory bodies adopt within their own jurisdictions. The noise limits are defined in ICAO Annex 16 and differ based on the rotorcraft size and configuration:

- MTOM > 3175 kg : ICAO Chapter 8;
- MTOM < 3175 kg : ICAO Chapter 8 *or* Chapter 11; and
- Tiltrotors : ICAO Chapter 13.

For novel configurations such as those we see for PAV or Urban Air Mobility (UAM), regulations do not currently exist and will be established as the technology matures with manufacturers and regulators working together to set out appropriate noise limits.

For conventional rotorcraft (MTOM> 3175 kg), the *ICAO Chapter 8* certification is defined at three predefined points. In particular, these are described as take-off, approach and flyover. On the other hand, for light helicopters certification under ICAO Chapter 11 is carried out at a single point—flyover. The manoeuvres for the three Chapter 8 certification points are shown in Fig. 6.2.

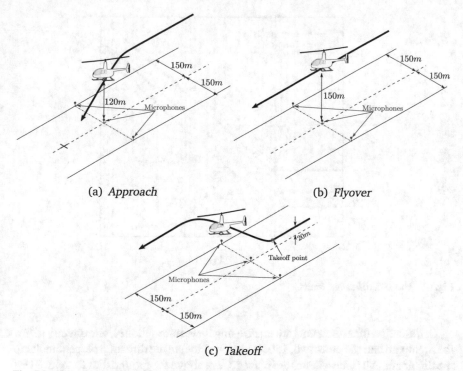

Fig. 6.2 Helicopter noise certification flight procedures

The approach conditions is set along the flightpath at a 6° glide slope, with the measurement taken at an altitude of 120 m. The test speed at the approach condition is carried out at either the speed corresponding to the best rate of climb or the minimum certified approach speed, whichever is greater. For the approach point, the rotorcraft's mass must lie between 90 and 105% of the Maximum Landing Mass.

For the flyover or overflight condition, the rotorcraft must fly a straight and level track at an altitude of 150 m. The test speed is based on the speed corresponding to maximum continuous power and the do not exceed speed. The test should be carried out between 90 and 105% of the rotorcraft's MTOM.

Finally, the takeoff procedures start at a 20 m level flight, 500 m from the microphones where takeoff power is applied and the rotorcraft commences a climb at an angle corresponding to the best rate of climb. During the test, the rotorcraft mass must lie between 90 and 105% of the MTOM. For all cases, the main rotor speed corresponds to the maximum RPM during normal operation. The manoeuvres for each certification point must be carried out within specified tolerances, with corrections provided for deviations from the reference procedures. This ensures that all rotorcraft have been evaluated at strictly the same points.

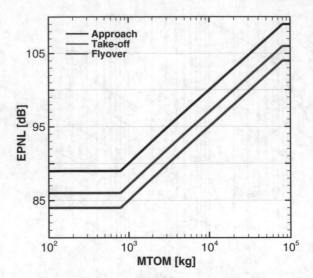

Fig. 6.3 Noise certification limits

Certification measurements are made using three microphones, one directly below the flightpath and the other two, 150 m on either side, in the direction perpendicular to the flightpath. All microphones are placed 1.2 m above the ground. For Chapter 8 certification, the Effective Perceived Noise Level (EPNL) metric is used to describe the noise during each manoeuvre, requiring microphone measurements at 0.5-s intervals. On the other hand, the *ICAO Chapter 11* rules are described by the Sound Exposure Level (SEL or L_{AE}).

Reference conditions for the measurements are at an ambient temperature of 25 °C, with a relative humidity of 70% with zero wind. Of course, real-world conditions will rarely reflect this, and as a result, regulators specify methods in which corrections should be made for deviations from reference conditions. However, there are upper limits to the deviations from the reference conditions, with corrections limited to 2 EPNLdB for flyover and approach conditions and 4 EPNLdB for take-off conditions.

Figure 6.3 presents the noise limits for the three certification points. Figure 6.3 shows that the upper noise limits are a function of the rotorcraft MTOM. In each case, the limits show a fixed value up to a MTOM of 800 kg after this, the noise levels increase at 3 dB for each doubling of the MTOM up until a MTOM of 80,000 kg, after which the noise limit is again constant.

Whilst the certification provides maximum permissible noise levels at the three operating points, the certification allows for some compromise. The noise levels at one or two of the specified points may exceed the stated maximums if it is offset at the other operating points. However, the compromise between points are subject to the following constraints:

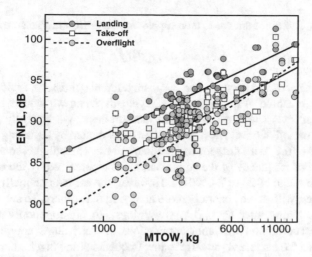

Fig. 6.4 Reported helicopter certification noise levels (*source* A. Filippone)

1. The sum of exceedance is not greater than 4 EPNLdB;
2. No single exceedance is greater than 3 EPNLdB; and
3. Exceedance can be offset at another certification point.

During certification, at least six valid measurements are required for each operating point. Typically more valid runs are recorded to satisfy the requirement that when all measurements are averaged, the error is within ±1.5 EPNLdB of the 90% confidence interval when all measurements (for all operating points) are averaged together.

The regulatory bodies record the noise limits for each rotorcraft that has sought certification, recent data for rotorcraft certification noise limits are shown in Fig. 6.4.

6.2.1 Noise Metrics

Noise is the perception of the physical pressure perturbations, with each individual experiencing the noise from a source differently. Therefore, the subjective nature of noise makes its impact difficult to quantify. In an effort to tackle this, a number of metrics have been developed that attempt to capture the various effects of a source and the perception this has on the ear. Many metrics have been developed over the many decades of acoustic research, we will only introduce those pertinent to the present discussion on rotorcraft noise.

Firstly, the most basic metric is the acoustic pressure and this is the physical mechanism that cause the vibration in the ear and the perception of a noise source. However, the ear has a very large dynamic range from around 20 μPa to 200 Pa. Working with values across this range is not feasible and typically a logarithmic scale is used. The noise source is then represented as a Sound Pressure Level:

$$SPL = 20 \log_{10}(p'/p_{ref}) \tag{6.1}$$

where p' is the acoustic pressure and p_{ref} is a reference pressure, typically $20\,\mu\text{Pa}$, which is the threshold of human hearing. Typically when we work with rotorcraft noise, we don't work with a single point measurement. We typically evaluate the sound over a period of time. The SPL can then be evaluated by replacing the acoustic pressure in Eq. (6.1) with the root-mean-square of the acoustic pressure signal.

The effect of frequency of the sound can also be perceived by the ear with perception in the range of 20 to 20,000 Hz. However, the ear is not equally sensitive to all frequencies, with some generally more annoying than others. To account for this, frequency weighting, such as A, B, C-weightings, etc. are commonly used, with A being the most common for aviation applications. A-weighting is applied to account for the fact that the ear is less sensitive to low frequencies and that higher frequencies are perceived as being more annoying. A-weighting is graphically shown in Fig. 6.5 and can be computed according to [2]:

$$L_A(f) = SPL(f) + 20 \log_{10}(G_A(f)) + 2 \text{ dB(A)}, \tag{6.2}$$

where G_A is the filter gain:

$$G_A(f) = \frac{12,194^2 f^4}{(20.6^2 + f^2)(107.7^2 + f^2)^{\frac{1}{2}}(737.9^2 + f^2)^{\frac{1}{2}}(12,194^2 + f^2)} \tag{6.3}$$

The Sound Exposure Level (SEL), also defined as L_{AE} is used in the certification of light rotorcraft (MTOM < 3175 kg). The SEL is used to represent all the acoustic energy measured over the event as if the duration of the event was just 1 s. In this way, the SEL captures the effects of the level and duration of the noise source. Furthermore, the SEL allows for sources of differing duration to be compared. The SEL can be calculated from noise measurements over a period of time or as the rotorcraft flies a particular trajectory. With reference to Fig. 6.6, the SEL can be computed from:

$$L_{AE} = 10 \log_{10} \left(\int_{t_1}^{t_2} 10^{SPLA(t)} dt \right) \tag{6.4}$$

where SPLA is the A-weighted SPL measured over the time period $t_1 \rightarrow t_2$, defined as the duration which the SPLA is above the reference value (L_{ref}). The reference value is 10 dB below the peak value (L_{max}). Note that the units for SEL are A-weighted decibels (dBA).

The EPNL metric is used in the certification of most rotorcraft (and also fixed-wing civil aircraft). The EPNL is a weighted integral metric with corrections for the tonal content and the effects of duration and loudness of the noise source. The EPNL

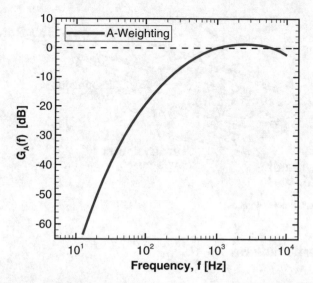

Fig. 6.5 A-weighting frequency gain

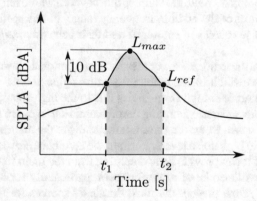

Fig. 6.6 Sound exposure level definition

metric is favoured for aircraft certification as it accounts for the annoyance, or human response, to a noise source. Like the SEL, the EPNL cannot be directly measured and requires computation over a noise event. However, the computation of the EPNL is a more laborious process and as a result, will not be presented here. Instead, the reader is directed towards Refs. [3, 4] which provide a thorough methodology for the EPNL calculation.

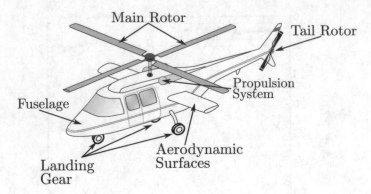

Fig. 6.7 Rotorcraft noise sources

6.3 Acoustic Sources

Much like the rotorcraft itself, the acoustics of the rotorcraft poses a serious problem. Rotorcraft noise comprises multiple sources that covers a whole range of directivities and spans across most of the audible frequency range. In this section, we will look at the dominant noise sources of rotorcraft and their behaviour as they propagate to the observer.

Figure 6.7 shows the major noise sources for a conventional rotorcraft. The propulsion system sources which includes the rotorcraft engine and the transmission are responsible for the noise at high frequencies due to the high rotational speed of the engine. The fuselage and corresponding aerodynamic surfaces are not typically significant sources of noise. However, their interaction with the main and tail rotors can subsequently lead to high noise levels. Further, the aerodynamic requirements of the fuselage mean that it can provide little shielding from the main rotor noise sources. The landing gear will also be of minor concern, particularly for low-speed events. The main and tail rotors present the most dominant sources for the rotorcraft and as a result will be the focus of the present section on rotorcraft noise sources. The sources of the main and tail rotor are similar in nature and can be further broken down into a number of additional sources. Whilst the sources for the main rotor will typically be found at the lower end of the frequency spectrum, the typically higher rotational speed and increased blade count of the tail rotor result in its sources being found higher up on the frequency spectrum. Therefore, the tail rotor noise can be subjectively more annoying than that of the main rotor.

The classification of the rotor sources has been the subject of great debate since the inception of the helicopter, and a great deal of studies have attempted to provide closure on the subject [5–8]. Rotor noise can be broadly grouped as harmonic noise and broadband noise, both of which can be described by a number of their own further groupings.

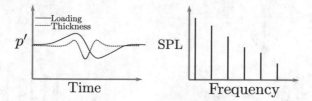

Fig. 6.8 Time and frequency response behaviour of steady loading and thickness sources

6.3.1 Harmonic Noise

Harmonic noise is periodic in nature and occurs at discrete frequencies attached to the rotor Blade Passage Frequency (BPF) and the frequency of the aerodynamic interaction. The BPF of a rotor is the product of the rotational frequency in revolutions per second (n) and the blade count (B):

$$\text{BPF} = nB \tag{6.5}$$

Steady Noise

Steady noise is driven by two mechanisms: the steady loading and the blade thickness. The first mechanism results from the mean aerodynamic loading, and in particular the thrust and torque (or lift and drag) produced by the rotor as it rotates. These forces appear periodic as the blade rotates with respect to a stationary observer. The second mechanism, the thickness source, results from the periodic displacement of the fluid as the blades move in space. Figure 6.8 shows the acoustic pressure of loading and thickness sources of a single blade over a revolution. When signals are combined for a multi-bladed rotor, the signals produce a sinusoidal response. The signal demonstrates that both mechanisms produce periodic signals in time that result in discrete pulses at integer multiples of BPF (harmonics) in the spectral space. Notably, the amplitudes of these sources rapidly decay with increasing harmonic.

The loading source is proportional to the blade loading. With the torque resulting in radiation in the rotor plane, and the thrust resulting in radiation normal to the rotor. However, as the thrust is typically greater in magnitude than the torque (per unit radius), the loading source is typically found radiating normal to the rotor. The thickness source is strongly affected by the rotor tip speed and radiates most effectively in the rotor plane.

Unsteady Loading

As a rotor blade traverses the azimuth it can be met with differing aerodynamic environments. As a result, the loading produced by the blade is not axisymmetric around the disc. This manifests itself as a periodic loading and subsequently periodic acoustic emissions. This is typically known as unsteady periodic or unsteady loading noise. The source is a result of unsteady loading on the blade and is tied to the frequency of the aerodynamic interaction.

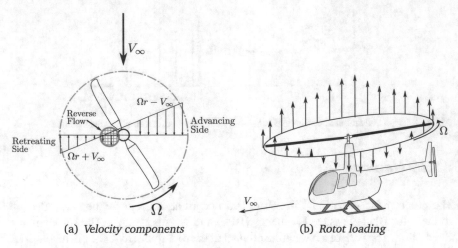

(a) *Velocity components*   (b) *Rotot loading*

Fig. 6.9 Unsteady loading due to forward flight

The unsteady loading source can occur throughout all phases of rotorcraft flight, including hover. During hover, interactions from the fuselage or operating in non-zero wind can result in an upwash and subsequently periodic loading on the rotor. Further, the periodic cyclic input required to trim the helicopter during hover can result in periodic loading on the rotor.

Whilst forward flight is an integral part of the rotorcraft operating corridor, it presents a very complex aeromechanics problem for rotorcraft. The resulting aerodynamic environment and controls required to trim the aircraft result in unsteady acoustic sources. Firstly, in forward flight the rotor sees differing velocity components as it rotates, Fig. 6.9a, with increased velocity on the advancing side and reduced velocity on the retreating side, and in some cases, regions of reversed flow. In order to alleviate the rolling moments that result from the velocity asymmetry from forward flight, rotorcraft have flexible blades that are allowed to flap. Flapping reduces the effective angle of attack on the advancing side, with the opposite effect on the retreating side. Further, similarly to hover, cyclic control is used in order to balance the moments about the rotor hub. The resultant effect of forward flight is a complex periodic loading around the disc Fig. 6.9b, with consequential acoustic emissions.

Unsteady loading also results from the interaction between rotorcraft components. For example, the interaction of the main rotor wake with the tail rotor or the interaction of the tail rotor with the hot exhaust gases from the propulsion system. Unsteady loading is particularly important for coaxial configurations, where the lower rotor is working directly below the wake of the top rotor.

Unsteady loading noise sources are characteristically very efficient radiators of sound. The resulting emissions are heard at frequencies corresponding to the frequency of the aerodynamic interaction and the rotor BPF. The resulting frequency spectrum, Fig. 6.10, no longer shows a rapid decay with increasing frequency as

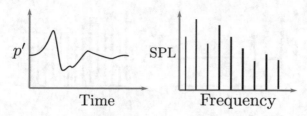

Fig. 6.10 Time and frequency response behaviour of unsteady loading sources

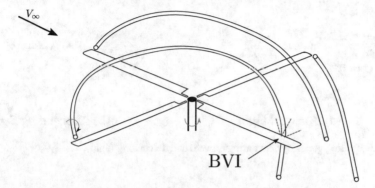

Fig. 6.11 Interaction of blade with tip vortex from previous blade resulting in BVI

we saw for the steady sources. Instead, the unsteady sources can persist to higher frequencies. In time, we no longer see the familiar sinusoidal pattern (for multiple blades), but the signal nonetheless remains periodic. The unsteady loading noise is also characterised as being highly directive. That is, high amplitudes persist in all directions away from the rotor. Therefore, the unsteady source is an important, and often dominant source in rotorcraft noise.

Blade Vortex Interaction

Blade Vortex Interaction (BVI) noise is a form of unsteady loading noise that is particularly important for rotorcraft. BVI occurs when a blade passes close to, or through, the tip vortex shed from a preceding rotor blade, Fig. 6.11. The interaction of the tip vortex with the rotor blade results in an impulse like loading at the leading edge. Consequently, there is a similarly impulsive acoustic emission that appears like sharp pulses, Fig. 6.12. Further, similar to unsteady noise, the amplitude does not rapidly decay and can persist to higher frequencies.

During forward flight and climbing flight, the wake trailed from the rotors is washed down and back away from the rotors. As a result, BVI is not of concern during these manoeuvres. However, during descent, the wake can sit just below the rotors or travel upwards and this is when BVI can occur. These two scenarios are illustrated in Fig. 6.13. Similarly to the general unsteady sources, BVI is an efficient radiator of sound across most directions. The strong directivity below the

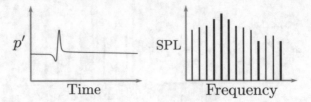

Fig. 6.12 BVI time and pressure histories for a single blade

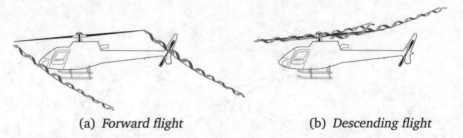

(a) *Forward flight*   (b) *Descending flight*

Fig. 6.13 Trailing vortex system during forward and descending flight

rotor plane, in addition to the occurrence during descent, BVI can pose significant challenges during certification at the approach manoeuvre.

BVI is very much characterised by the vortex it interacts with, for example, its size, strength and orientation (see for example [9]). It is also strongly affected by the rotor tip shape, Brocklehurst and Barakos [10] presented a review of tip rotor shapes and demonstrated their modification to reduce BVI noise. BVI is a very active area of rotorcraft research and Lowson [11] and Yu [12] present a much more thorough review of the subject.

High-Speed Impulsive Noise (HSI)

High Speed Impulsive (HSI) noise is a form of thickness noise that occurs due to compressibility effects as blade sections approach transonic speeds. HSI is particularly important during forward flight manoeuvres when the combined effects of the rotorcraft's speed and the rotor rotational speed result in high relative velocities on the advancing blade.

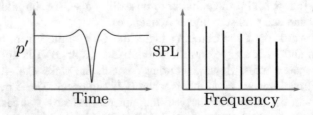

Fig. 6.14 HSI time and pressure histories for a single blade

HSI is similar in character to BVI noise in that it produces a very sharp impulsive acoustic emission, Figure 6.14. Like the thickness source that it originates from, HSI is an effective radiator in the rotor plane. HSI can be a significant source during forward flight and as a result, can pose challenges for the flyover certification point.

HSI is closely associated with transonic relative speeds and the appearance of shocks over the blade surface. Therefore, HSI is closely coupled to the rotor tip speed and reducing tip speed can reduce the effects or onset of HSI. Further, the blade thickness can play an important role in HSI with modern blades designed with thin blade tips. Modern blades also feature swept-back tips which help reduce the relative velocity seen by each blade element and can consequently reduce the onset of HSI.

6.3.2   Broadband Noise

Broadband noise, sometimes known as vortex noise, is characterised by random pressure fluctuations on the blade surface. These random fluctuations in blade loading result from interaction with turbulent flow. Broadband noise is therefore a stochastic source that appears as loading noise.

Broadband noise is characterised as random pressure fluctuations in the time domain which appears as a continuous spectrum in the frequency domain, Fig. 6.15. Broadband noise is typically dominant at higher frequencies where the tonal component has decayed. This is particularly important where the main rotor has a low BPF and the most dominant tones are below the audible frequency range. When the rotorcraft sources are corrected for frequency (for example using A-weighting), the broadband source is found to be the most important. Further, the high-frequency component results in the broadband sources being the most subjectively annoying sources. However, the propagation of sound is much less efficient with increasing frequency. Therefore, the broadband source is less important for far-field noise, i.e. where the rotorcraft is far from the observer. On the other hand, it is important in the nearfield, particularly overhead, as the dominant directivity lies below the rotor plane.

There are a large number of physical mechanisms that generate broadband noise and are broadly characterised as either self noise or turbulent ingestion noise.

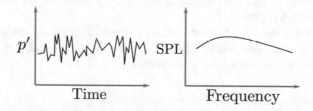

Fig. 6.15 Broadband time and pressure histories

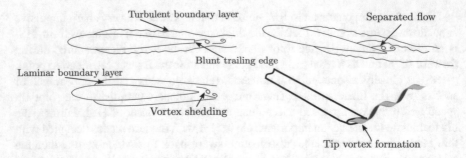

Fig. 6.16 Broadband noise sources

Self Noise

The self noise group of broadband sources are linked to turbulence generated by the rotor blade itself. The self noise is closely associated with the properties of the boundary layer over the blade surface. Turbulent boundary layers are inherently coupled to broadband noise. Of particular importance to the broadband noise is the turbulence scattering at the blade trailing edge, particularly blunt trailing edges. A further source arises when the boundary layer separates from the blade surface. Even for laminar boundary layers, the vortex shedding will result in broadband noise sources. Finally of note is the sources that result from the formation of the tip vortices at the blade tips. These sources are summarised in Fig. 6.16. The self noise broadband sources are characterised as being at the high-frequency range of the spectrum and whilst they can be more subjectively annoying, they are less important when the observer is in the far field.

Turbulent Ingestion Noise

Turbulent ingestion noise is associated with the interaction of the blade with an externally generated source of turbulence. One particularly important source is Blade Wake Interaction (BWI) noise. This results when a blade passes through a wake system of a preceding blade. BWI can also be generated as the tail rotor interacts with the wake from the main rotor. BWI is also significant where the rotor is near the ground due to recirculation. One special case of BWI arises in the case of co-axial or overlapping rotors where the bottom rotor operates within the wake of the top rotor. In this case, the bottom rotor operates almost entirely with some level of turbulence ingestion. Turbulent ingestion noise also occurs when the blade interacts with atmospheric turbulence.

Turbulent ingestion noise is characterised in the mid-frequency range so may be less subjectively annoying when corrected for frequency but will propagate more efficiently than the self noise broadband sources.

6.4 Acoustic Modelling

Having discussed the range of rotorcraft noise sources, we now turn to modelling these sources. Rotorcraft noise prediction has advanced tremendously over the past few decades—mostly in part to the advances in computing. Further, the improvements in aerodynamic prediction have helped advance accurate acoustic predictions. It is important to note that the ability to predict accurate rotorcraft noise, not only relies on the noise modelling, but also on the quality of the aerodynamic data.

The earliest works on rotor noise were focused on propeller noise, with Gutin laying down one of the first theoretical developments for harmonic rotor noise due to loading [13]. Subsequent advancements for thickness sources and the effects of forward flight were made. However, due to the limited development of the rotorcraft itself, work was all focused on propellers. It took until the 1960s, when rotorcraft came into their own, both as a commercial vehicle and a military machine, until rotorcraft noise started to receive the required attention. For the reader's own interest, the historical developments of rotorcraft noise prediction is presented by Brentner and Farassat [7]. One of the most significant developments in rotorcraft noise prediction was made in 1969 when Ffowcs Williams and Hawkings published their work [14] which would set the groundwork for future rotorcraft noise codes. Ffowcs Williams and Hawkings presented a generalization of Lighthill's acoustic analogy to surfaces in arbitrary motion [15] which offered significant advances in acoustic modelling for rotorcraft. Many developments in acoustic codes have been made since. However, most are based on the work of Lighthill and subsequently Ffowcs Williams and Hawkings.

To proceed in this section, we will look closer at the development of Lighthill's acoustic analogy and derive the Ffowcs Williams and Hawkings (FWH) equation, looking closely at the physical description of noise that it provides. Following this, we will look at further developments of these equations in application to modelling of the rotorcraft sources that we discussed in the previous section. We will also present a numerical implementation and sample MATLAB code for the thickness and loading sources, see Appendix A.

Lighthill's Acoustic Analogy

Lighthill was concerned primarily with turbojet engine noise. However, Lighthill derived general expressions relating the aerodynamically generated noise. Starting from the Navier-Stokes equations, the conservation of mass and momentum are:

$$\frac{\partial \rho}{\partial t} + \frac{\partial \rho u_j}{\partial x_j} = 0 \tag{6.6}$$

$$\rho \frac{\partial u_i}{\partial t} + \rho u_j \frac{\partial u_i}{\partial x_j} + \frac{\partial p}{\partial x_i} - \frac{\partial \tau_{ij}}{\partial x_j} = 0 \tag{6.7}$$

where ρ is the density; p the pressure; u_i the velocity; and τ_{ij} the viscous stress tensor:

$$\tau_{ij} = \mu \left[\frac{\partial u_i}{\partial x_j} \frac{\partial u_j}{\partial x_i} - \frac{2}{3} \delta_{ij} \frac{\partial u_k}{\partial x_k} \right] \tag{6.8}$$

To derive Lighthill's equation we must subtract the time derivative of continuity with the divergence of the conservation of mass. The time derivative of the continuity equation is:

$$\frac{\partial}{\partial t} \left(\frac{\partial \rho}{\partial t} + \frac{\partial \rho u_i}{\partial x_i} \right) = \frac{\partial^2 \rho}{\partial t^2} + \frac{\partial}{\partial t} \frac{\partial \rho u_i}{\partial x_i} = 0 \tag{6.9}$$

and the divergence of the conservation of momentum:

$$\frac{\partial}{\partial x_i} \left(\rho \frac{\partial u_i}{\partial t} + \rho u_j \frac{\partial u_i}{\partial x_j} + \frac{\partial p}{\partial x_i} - \frac{\partial \tau_{ij}}{\partial x_j} \right) = \frac{\partial}{\partial x_i} \frac{\partial}{\partial t} \rho u_i$$

$$+ \frac{\partial^2}{\partial x_i \partial x_j} \left(\rho u_i u_j + p \delta_{ij} - \tau_{ij} \right) = 0 \tag{6.10}$$

upon subtraction we have:

$$\frac{\partial^2 \rho}{\partial t^2} = \frac{\partial^2}{\partial x_i \partial x_j} \left(\rho u_i u_j + p \delta_{ij} - \tau_{ij} \right) \tag{6.11}$$

Lighthill defined a stress tensor:

$$T_{ij} = \rho u_i u_j + \delta_{ij} (p - \rho c_0{}^2 \rho) - \tau_{ij} \tag{6.12}$$

where c_0 is the local sound speed. Upon substitution we need to subtract $c_0{}^2 \frac{\partial^2 \rho}{\partial x_i{}^2}$ from each side to give:

$$\frac{\partial^2 \rho}{\partial t^2} - c_0{}^2 \frac{\partial^2 \rho}{\partial x_i{}^2} = \frac{\partial^2 T_{ij}}{\partial x_i \partial x_j} \tag{6.13}$$

Introducing the density perturbation: $\rho' = \rho - \rho_\infty$:

$$\frac{\partial^2 \rho'}{\partial t^2} - c_0{}^2 \frac{\partial^2 \rho'}{\partial x_i{}^2} = \frac{\partial^2 T_{ij}}{\partial x_i \partial x_j} \tag{6.14}$$

Note that $\frac{\partial \rho}{\partial t} = \frac{\partial \rho'}{\partial t}$ and that Lighthill's stress tensor is now: $T_{ij} = \rho u_i u_j + \delta_{ij} (p' - c_0{}^2 \rho') - \tau_{ij}$. This is Lighthill's acoustic analogy, a re-arrangement of the Navier-Stokes equations that is exact and without approximation into an acoustic wave equation. The left hand side of the equation defines the sound propagation at c_0 through a uniform medium, whilst the right hand side describes the source terms.

For small perturbations, the acoustic density and pressure are related:

$$p' = c_0^2 \rho'$$ (6.15)

we can then represent the acoustic sources in terms of the more familiar acoustic pressure:

$$\Box^2 p' = \frac{\partial^2 T_{ij}}{\partial x_i \partial x_j}$$ (6.16)

where $\Box^2 = \frac{1}{c_0^2}\frac{\partial^2}{\partial t^2} - \frac{\partial^2}{\partial x_i \partial x_j}$ is the wave operator.

The Ffowcs Williams and Hawkings Equation

Ffowcs Williams and Hawkings derived expressions which expanded Lighthill's work to include the effects of arbitrary surfaces in motion. This approach proved vital in the development of propeller and rotorcraft codes which inherently rely on the movement of aerodynamic surfaces to generate useful forces.

The derivation of the FWH equations begins by considering an unbounded fluid domain discretised into mathematical surfaces representing the moving surface. The fluid outside the surface is equivalent to the real fluid flow, whilst inside the surface the fluid is described by the undisturbed medium. This results in discontinuities on the surface and this presents a problems in evaluating the derivatives within the conservation equations. To handle these discontinuities, Ffwocs Wiliams and Hawkings utilised generalised function theory.

Generalised function theory can be a challenging subject to grasp and apply. Further, it is beyond the present scope to go into any great detail on the subject of generalised functions and the reader is directed to Farassats work on the subject [16]. However, to proceed with the derivation, we must introduce some preliminary topics in generalised function theory and we will follow the example discussed by Brentner and Farassat [8]. First, if we consider some arbitrary function, $p(x)$ that is piecewise smooth, with a discontinuity at $x = x_0$, the generalised derivative is:

$$\frac{\bar{d}p}{dx} = \bar{p}'(x) = p'(x) + \Delta p \delta(x - x_0)$$ (6.17)

where $p'(x)$ is the derivative of the function $p(x)$ and δ is the Dirac delta function. To continue, we now consider a function, $q(\vec{x})$, where $\vec{x} = (x_1, x_2, x_3)$ that has a discontinuity across the surface $f(\vec{x}) = 0$. If we now define a 'jump' Δq in the function $q(\vec{x})$ across the surface $f(\vec{x}) = 0$, such that:

$$\Delta q = q(f = 0^+) - q(f = 0-)$$ (6.18)

the generalised partial derivative is:

$$\frac{\bar{\partial} q}{\partial x_i} = \frac{\partial q}{\partial x_i} + \Delta q \frac{\partial f}{\partial x_i} \delta(f) \tag{6.19}$$

With this little knowledge on generalised function theory, we can continue to our problem of the moving surface. Let us define a moving surface which contains within it all the noise sources, $f(\vec{x}, t) = 0$, Fig. 6.17. Outside the surface $f(\vec{x}, t) > 0$, whilst inside $f(\vec{x}, t) < 0$. Further, the surface normal is defined as: $\nabla f = \vec{n}$, which is outwards positive. Our goal is to create an inhomogeneous wave equation derived from the Navier-Stokes equation. To proceed as with Lighthill, we must use generalised function theory to accommodate the discontinuity at the surface. As described by Bretner and Farassat [8], we assume that the fluid extends into the surface and here has the properties of the undisturbed quiescent fluid. This results in the discontinuity at the surface and the approach allows us to utilise the Green function in order to solve the acoustic problem over the entire domain. Assuming no further discontinuities in the fluid domain, we begin with the equations for the conservation of mass and momentum in the region outwith the surface, Eqs. (6.6) and (6.7). With our discontinuity at the surface, using the theory of generalised functions, the conservation of mass is now:

$$\frac{\bar{\partial} \rho}{\partial t} + \frac{\bar{\partial}}{\partial x_i}(\rho u_i) = \frac{\partial \rho}{\partial t} + \Delta \rho \frac{\partial f}{\partial t} \delta(f) + \frac{\partial}{\partial x_i}(\rho u_i) + \Delta(\rho u_i)\frac{\partial f}{\partial x_i}\delta(f) \tag{6.20}$$

similarly for the conservation of momentum:

$$\frac{\bar{\partial}}{\partial t}(\rho u_i) + \frac{\bar{\partial}}{\partial x_j}\left(\rho u_i u_j + p\delta_{ij} - \tau_{ij}\right) = \frac{\partial}{\partial t}(\rho u_i) + \Delta(\rho u_i)\frac{\partial f}{\partial t}\delta(f) \tag{6.21}$$

$$+ \frac{\partial}{\partial x_j}\left(\rho u_i u_j + p\delta_{ij} - \tau_{ij}\right)$$

$$+ \Delta\left(\rho u_i u_j + p\delta_{ij} - \tau_{ij}\right)\frac{\partial f}{\partial x_i}\delta(f)$$

If we recall that the unit normal $\vec{n} = \frac{\partial f}{\partial x_i}$, we define the local fluid velocity in the normal direction: $u_n = u_i n_i$ and the local velocity of the surface in the normal direction: $v_n = -\frac{\partial f}{\partial t}$. Further, if we substitute the 'jump' at the surface: $\Delta \rho = \rho - \rho_0$; $\Delta \rho u_i = \rho u_i$; $\Delta\left(\rho u_i u_j + p\delta_{ij} - \tau_{ij}\right) = \left(\rho u_i u_j + p\delta_{ij} - \tau_{ij}\right)$, where the subscript zero refers to the properties of the undisturbed quiescent fluid. We therefore have for the conservation of mass:

$$\frac{\bar{\partial} \rho}{\partial t} + \frac{\bar{\partial}}{\partial x_i}(\rho u_i) = \frac{\partial \rho}{\partial t} + v_n(\rho - \rho_0)\delta(f) + \frac{\partial}{\partial x_i}(\rho u_i) + \rho u_n \delta(f) \tag{6.22}$$

noting from Eq. (6.6) that $\frac{\partial \rho}{\partial t} = -\frac{\partial \rho u_j}{\partial x_j}$:

$$\frac{\bar{\partial}\rho}{\partial t} + \frac{\bar{\partial}}{\partial x_i}(\rho u_i) = [\rho_0 v_n + \rho(u_n - v_n)]\delta(f) \tag{6.23}$$

performing a similar operation on the conservation of momentum:

$$\frac{\bar{\partial}}{\partial t}(\rho u_i) + \frac{\bar{\partial}}{\partial x_j}\left(\rho u_i u_j + p\delta_{ij} - \tau_{ij}\right) = \frac{\partial}{\partial t}(\rho u_i) + \rho u_i v_n \delta(f) \tag{6.24}$$

$$+ \frac{\partial}{\partial x_j}\left(\rho u_i u_j + p\delta_{ij} - \tau_{ij}\right)$$

$$+ \left(\rho u_i u_j + p\delta_{ij} - \tau_{ij}\right)n_j\delta(f)$$

and noting from Eq. (6.7) that $\rho\frac{\partial u_i}{\partial t} = -\frac{\partial}{\partial x_j}(\rho u_i u_j + p\delta_{ij} - \tau_{ij})$, we have:

$$\frac{\bar{\partial}}{\partial t}(\rho u_i) + \frac{\bar{\partial}}{\partial x_j}\left(\rho u_i u_j + p\delta_{ij} - \tau_{ij}\right) = \left[\rho u_i(u_n - v_n) + (p\delta_{ij} - \tau_{ij})n_j\right]\delta(f) \tag{6.25}$$

To proceed with the derivation, we carry out the same procedure outlined in the derivation of Lighthill's acoustic analogy. Namely, we subtract the time derivative of the conservation of mass, Eq. (6.23), from the divergence of the conservation of momentum, Eq. (6.25):

$$\frac{\bar{\partial}^2\rho}{\partial t} = \frac{\bar{\partial}}{\partial t}\{[\rho_0 v_n + \rho(u_n - v_n)]\delta(f)\} \tag{6.26}$$

$$- \frac{\bar{\partial}}{\partial x_i}\{[\rho u_i(u_n - v_n) + (p\delta_{ij} - \tau_{ij})n_j]\delta(f)\}$$

$$+ \frac{\bar{\partial}^2}{\partial x_i \partial x_j}\left(\rho u_i u_j + p\delta_{ij} - \tau_{ij}\right)$$

substituting for Lighthill's stress tensor, T_{ij}:

$$\frac{\bar{\partial}^2\rho'}{\partial t} - c_0{}^2\frac{\bar{\partial}^2\rho}{\partial x_i{}^2} = \frac{\bar{\partial}}{\partial t}\{[\rho_0 v_n + \rho(u_n - v_n)]\delta(f)\} \tag{6.27}$$

$$- \frac{\bar{\partial}}{\partial x_i}\{[\rho u_i(u_n - v_n) + (p\delta_{ij} - \tau_{ij})n_j]\delta(f)\}$$

$$+ \frac{\bar{\partial}^2}{\partial x_i \partial x_j}\left(T_{ij}H(f)\right)$$

This is the FWH equation which is a re-arrangement of the Navier-Stokes equation as an inhomogeneous wave equation. Note that following Ffowcs Williams and Hawkings, we have introduced the heaviside function $H(f)$ to emphasize that the

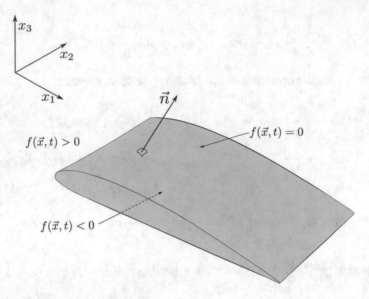

Fig. 6.17 Definition of the moving surface

final term is only applicable outside of the surface (interior $T_{ij} = 0$). Finally, we can transform from the acoustic density to the acoustic pressure as was done previously and neglecting the viscous source which is often negligible in comparison with the other sources, we have the more familiar form of the FWH equation:

$$\bar{\Box}^2 p'(\vec{x}, t) = \frac{\bar{\partial}}{\partial t} \{[\rho_0 v_n + \rho(u_n - v_n)]\,\delta(f)\} - \frac{\bar{\partial}}{\partial x_i} \{[p n_i + \rho u_i (u_n - v_n)]\,\delta(f)\} + \frac{\bar{\partial}^2}{\partial x_i \partial x_j} [T_{ij} H(f)]$$
(6.28)

The above equation is a powerful equation that describes the acoustic emissions of an arbitrary surface in motion and holds for permeable or impermeable surfaces. In the case of a impermeable surface (e.g. the rotor blade is the integration surface), the normal velocity of the surface and the fluid at the surface are equal, and we substitute $v_n = u_n$.

One of the great advantages of the Ffowcs Williams Hawkings equation is its ability to physically describe the sources. The right hand side is comprised of three components. The first term on the right hand side is the monopole source and describes the thickness noise. The second term is the dipole source and describes the loading sources. The final term on the right hand side is known as the quadrupole or volume source. It is important when the blade approaches transonic speeds and requires detailed knowledge of the flow off of the solid surface. Due to the increased computational cost associated with volume integration, the quadrupole source is often neglected unless the blade is operating in the transonic regime.

In their fundamental work, Ffowcs Williams and Hawkings [14] also derived the governing equations of the Kirchhoff formulation, based on the generalised wave equation. The Kirchhoff formulation represents an arbitrary fictitious surface (the Kirchhoff surface), enclosing all of the physical sources. The use of the Kirchhoff surface avoids the need for volume integration so can be particularly attractive for application on high-speed rotors. However, detailed flow-field calculations are still required. Further, placement of the Kirchhoff surface is not a trivial task, requiring sufficient volume to capture of acoustic sources, whilst minimising computational cost (which arises due to the need for high temporal and spatial resolution of the flow-field). Finally, the use of the Ffowcs Williams Hawkings equation as a porous surface (similar to the Kirchhoff formulation) to handle quadrupole sources has been shown to work extremely well for rotorcraft problems. Brentner and Farassat [17] present a detailed comparison of both Ffowcs William Hawkings and Kirchhoff methods. In the next sections, we will focus on developments of acoustic methods based on the Ffowcs Williams and Hawkings as this represents the majority of acoustic codes in the literature.

In recent years, the improvements in computational power has increased the capabilities of aerodynamic solvers, allowing engineers to directly solve the flowfield in terms of acoustic pressure. However, this approach is still prohibitively expensive, given the requirements for high spatio-temporal resolution, and as a result, it is only currently appropriate for the nearfield analysis. As an alternative, we can use an integral formulation.

The FWH equation, as an inhomogenous wave equation, can be solved using the Green function for free space. This allows us to separate the aerodynamics from the acoustics, where the aerodynamics is an input to the acoustic solution. This is known as an "integral formulation" and provides a great advantage to rotorcraft acoustics. Our computational domain can now be reduced significantly, with dependence on the aerodynamic sources. Further, neglecting the quadrupole source we then require only a surface integration and we can obtain solutions to the acoustic problems from low-order methods. A number of integral formulations have been presented in the literature. Perhaps the most well known and widely applied is based on the work of Farassat, which we utilise in the proceeding section.

6.4.1 Thickness and Loading

From the FWH equation, one can develop a solution either in the frequency domain or the time domain. Both approaches have been shown capable of predicting rotorcraft noise. However, for rotorcraft noise in particular, Bretner [18, 19] has shown that the blade kinematics plays an important part of the rotorcraft noise emissions. Typically as the blade kinematics and the aerodynamics, either from CFD or some other approach will be reported with respect to time (or blade azimuth), we will proceed with modelling the thickness and loading sources in the time domain. In particular, we will follow the formulation of Farassat as this represents the most widely applied approach for computing the thickness and loading sources of rotorcraft.

Neglecting the quadrupole source in the FWH equation, Eq. (6.28), leaves thickness and loading sources. If the rotor speed is not in the transonic regime, this offers a much simplified evaluation of the rotor noise sources. Further, assuming an impenetrable surface (i.e. $v_n = u_n$), we now have the following equations to solve:

$$\bar{\Box}^2 p'_T(\vec{x}, t) = \frac{\bar{\partial}}{\partial t} [\rho_0 v_n \delta(f)] \tag{6.29}$$

$$\bar{\Box}^2 p'_L(\vec{x}, t) = -\frac{\bar{\partial}}{\partial x_i} [p n_i \delta(f)] \tag{6.30}$$

where p'_T and p'_L are the thickness and loading noise sources respectively.

Farassat used a retarded time formulation and the free-field Green's function to derive solutions for the thickness and loading sources above. First deriving Formulation 1 [20, 21] and subsequently Formulation 1A [22], Farassat's work has been widely used in the prediction of rotor noise. From Farassat, the acoustic pressure is given as a sum of the thickness and loading sources:

$$p'(\boldsymbol{x}, t) = p_L'(\boldsymbol{x}, t) + p_T'(\boldsymbol{x}, t) \tag{6.31}$$

where

$$4\pi p_T'(\boldsymbol{x}, t) = \rho c_0 \int_{f=0} \left[\frac{\dot{M}_n}{\mathscr{R}(1 - M_r)^2} + \frac{M_n \dot{M}_r}{\mathscr{R}(1 - M_r)^3} \right]_{ret} \mathrm{d}S$$
$$+ \rho c_0 \int_{f=0} \left[\frac{c_0 M_n (M_r - M^2)}{\mathscr{R}^2 (1 - M_r)^3} \right]_{ret} \mathrm{d}S \tag{6.32}$$

and

$$4\pi p_L'(\boldsymbol{x}, t) = \int_{f=0} \left[\frac{\dot{p}_S \cos\Theta}{c_0 \mathscr{R}(1 - M_r)^2} + \frac{\dot{M}_r p_S \cos\Theta}{\mathscr{R}(1 - M_r)^3} \right]_{ret} \mathrm{d}S$$
$$+ \int_{f=0} \left[\frac{p_S(\cos\Theta - M_n)}{\mathscr{R}^2 (1 - M_r)^2} + \frac{p_S \cos\Theta (M_r - M^2)}{\mathscr{R}^2 (1 - M_r)^3} \right]_{ret} \mathrm{d}S \tag{6.33}$$

p_S is the blade surface pressure from the aerodynamic loading on the blade; M_r and M_n are the projection of the local Mach number, M, in the observer and normal directions respectively; Θ is the angle between the radiation and normal directions; $\mathscr{R}$ is the distance between source and observer. The $|_{ret}$ subscript indicates that the integrals are evaluated at the source time (or retarded time). The $1/\mathscr{R}$ terms represent the far-field contribution whilst the $1/\mathscr{R}^2$ terms represent the nearfield contributions.

The above equations are an integral formulation of the FWH equations for loading and thickness noise. Therefore, the aerodynamic solution is required as input in order to solve the acoustic problem. Further, we require the blade motion, in addition to it's derivatives, at each point in source time. This is an advantageous approach where the aerodynamics is solved separately (with minimal constraints due to acoustics) to the acoustics. Further, the blade kinematics are also typically evaluated during the aerodynamic evaluation. As a result, the acoustic code can be completely separate and optimised for the acoustic computation only. The aerodynamics can be evaluated using any approach, but in the above expression, the blade surface pressure is required which will typically necessitate the use of CFD computations to achieve accurate predictions.

6.4.2 Compact Chord Formulation

During preliminary design or parametric studies of rotor blades, low-order methods such as Blade Element Momentum Theory are often utilised. In this case, we do not have the blade surface pressure, but instead point loading at a number of blade spanwise panels. The variation in loading can be assumed compact if the chordwise variation in retarded time is negligible. Bretner and Jones [23] showed that the loading noise from a chordwise compact source is:

$$
4\pi p_L'(\mathbf{x}, t) = \int_{f=0} \left[\frac{\dot{l}_r}{c_0 \mathscr{R}(1 - M_r)^2} + \frac{l_r \dot{M}_r}{\mathscr{R}(1 - M_r)^3} \right]_{ret} dS
$$
$$
+ \int_{f=0} \left[\frac{l_r - l_M}{\mathscr{R}^2(1 - M_r)^2} + \frac{l_r(M_r - M^2)}{\mathscr{R}^2(1 - M_r)^3} \right]_{ret} dS
$$

(6.34)

where $\vec{l}$ is now the blade element loading vector. The chordwise compact assumption has been used widely, particularly where blade surface pressure data is not available and has been shown to be generally accurate for observer locations away from the tip path plane [19]. The chordwise compact assumption applies only to the loading source. For the thickness source, Lopes [24] has derived a compact chord expression based on the thickness term of Formulation 1A assuming that the radiation distance is much greater than that of the blade element chord.

6.4.3 Numerical Implementation

There are a number of approaches to implementing Formulation 1A of Farassat as the reader may find by reviewing the literature. Here we will outline some of the significant choices to be made when developing and implementing Formulation 1A within an acoustic code.

Formulation 1A of Farassat is a retarded-time formulation where the surface integration is evaluated at the 'retarded-time' : $\tau = t - r/c_0$. Where the surface is a discretisation of the rotor blade into a number of panels representing acoustic sources. The retarded time formulation can be evaluated following one of two approaches. In particular, we can place the emphasis on the observer: observer dominant algorithm; or place the emphasis on the source: source dominant algorithm.

In the observer dominant algorithm, the observer time is selected and the emission time of each source and it's location in space at that time is computed. The approach results in unequally spaced emission time for each panel. As we are considering rotor blades which move in time and space, a system of equations must be solved to compute the source time and location. However, this can easily be implemented within a simple Newton-Raphson solver.

In the source dominant algorithm, the source time is specified and the corresponding time that the source reaches the observer is evaluated. The sources will reach the observer at a time $t = \tau + r_i/c_0$. In this approach the sources will reach the observer at different times and as such a 'common' observer time is specified and the actual time the sources reach the observer are interpolated onto this common time array. This approach may be initially more intuitive as the source locations are known (e.g. through the aerodynamic calculation) at the source time and do not have to be computed. Brentner et al. compared both observer dominant and source dominant algorithms [25] and found that they performed equally well for general rotor blade cases. However, for cases such as manoeuvring flight the source dominant algorithm was found to require significantly less computational operations in the noise evaluation.

The numerical implementation of Formulation 1A of Farassat can be further implemented based on the approximation of the integration. If we consider the generic retarded time formulation, where the acoustic pressure of the observer at time t located at $\vec{x}$ is proportional to the source strength, Q located at $\vec{y}$ at source time τ:

$$p'(\vec{x}, t) = \frac{1}{4\pi} \int_{f=0} \left[\frac{Q(\vec{y}, \tau)}{r|1 - M_r|} \right]_{ret} dS \qquad (6.35)$$

In the mid-panel quadrature approach, the blade surface is divided into N_p panels in the spanwise and chordwise directions. The integrals are evaluated based on the properties at the mid-point of each panel and at the source time, which is selected dependent on whether the source dominant or observer dominant implementations are utilised. The acoustic pressure is then:

$$p'(\vec{x}, t) \approx \frac{1}{4\pi} \sum_{i=1}^{N_p} \left[\frac{Q(\vec{y}, t - r_i/c_0)}{r_i|1 - M_r|_i} \right]_{ret} \Delta S_i \qquad (6.36)$$

This approach forms the basis of most acoustic predictions and due to the simplicity of computing mid-panel properties from given panel data, is easily implemented. When selecting this approach, it is important to ensure that the panels are sufficiently small such that the source properties do not change significantly over the panel and that changes in retarded time over the panel is negligible. In the case that mid-panel quadrature provides insufficient accuracy, alternative methods offering improved accuracy are available, see Brentner [26].

6.4.4   Blade-Vortex Interaction Noise (BVI)

Blade Vortex Interaction (BVI) noise, as we have previously discussed, is a form of unsteady loading noise that results from interaction of the blade with a tip vortex from a preceding blade. Therefore, our existing methods are capable of predicting this source. However, the BVI source presents some further complications. Due to the impulsive nature of BVI, accurate aerodynamic data of sufficiently small time step is required to afford accurate acoustic prediction. However, accurate aerodynamic data can be challenging and must resolve the main contributors to BVI noise such as the vortex strength, orientation and its transient interaction with the blade. Nonetheless, aerodynamic predictions are presently mature enough to afford us the ability to accurately predict BVI. Furthermore, as a significant noise source, large efforts are currently ongoing to better understand and reduce this source.

6.4.5   HSI

High Speed Impulsive (HSI) noise becomes important where regions of the rotor blade see transonic velocities. Therefore, the computation of HSI noise requires the evaluation of the quadrupole source in the FWH equation. This is typically a computationally expensive task, requiring knowledge of the flow off the blade. An alternative would be to use a Kirchhoff formulation or a porous FWH formulation which can avoid the volume integral evaluation. In this instance, we place an arbitrary surface at some location off of the blade surface. Within the arbitrary surface, we make direct computation of the acoustic variables. Typically, a high mesh density is required within the surface in order to resolve the various turbulent structures as well as acoustic frequency resolution. Therefore, the placement of the surface is a difficult task, it should be far enough away from the surface to capture the noise sources of interest whilst being as close to the surface as possible to reduce the computational cost. HSI noise is dominant in the rotor plane and vanishes rapidly away from this location. Therefore, some resources can be saved by including this source only for observers in this location and neglecting the computation otherwise.

6.4.6 Broadband Noise

Broadband noise is the result of random pressure fluctuations on the blade surface caused by turbulence. As a result, it can be characterised by the loading term in the FWH equation. However, its evaluation is not so simple, due to the difficulty of computing the turbulent flow features and subsequently the resulting loading on the blade surface. Therefore, computation of the broadband noise typically follows a statistical approach and with a number of semi-analytical components. Broadband noise is typically evaluated in the frequency domain, although it is worth noting that Casper and Farassat [27] have derived expressions to evaluate broadband noise in the time domain that corresponds well to computations in the frequency domain.

A further difficulty with broadband noise sources is that there are a large number of sources that contribute to broadband noise and discerning specific sources can be challenging when analysing noise signals. It also adds the additional challenge of modelling a significant number of sources. Brooks and Burley [28] present an overview of modelling both self noise and turbulence ingestion broadband sources. However, as is common with broadband noise modelling, the expressions rely on properties of the turbulence and the boundary layer response. Amiet [29, 30] derived a number of semi-empirical expressions for boundary layer properties specifically for the predictions of broadband noise. The expressions of Amiet are restricted to flow past an aerofoil. Nonetheless, they can reasonably predict various broadband sources from rotors and work extremely well within a strip theory approach.

Pegg [31] presented a semi-empirical expression for rotor broadband noise based on the blade loading, tip speed and the blade area. Whilst the expression is very general and not representative of a specific broadband source, it still serves as a useful prediction tool for preliminary analyses. Brooks et al. [32] also presented semi-empirical relations, however, focused on specific self noise sources. They also presented comparisons with experimental data in order to compare the relations. Their expressions were developed on scaled boundary layer properties from measured data. Again, this approach would be appropriate for preliminary studies in order to account for the effects of broadband noise.

Broadband noise is typically high-frequency. Therefore, it can subjectively be the most annoying source, particularly when the noise is corrected for frequency. Further, the source will be more important in the nearfield and as a result should not be neglected when considering modelling of certification procedures. However, in the farfield, the broadband source will be less important due to the greater attenuation before it reaches the observer.

6.5 Noise Reduction

Noise reduction is presently the focus of much research due to the reasons outlined in the opening section of this chapter. A number of design choices can be made to reduce the noise in terms of both geometrical and operational parameters. However, perhaps the most important tool an engineer can have is to have a better understanding of rotorcraft noise and its sources. With a better understanding of the sources, the engineer can best apply their energy to target the most important noise reductions for a particular application.

Besides addressing the noise from a greater understanding, there are a number of established approaches for reducing rotorcraft noise. Firstly, we can reduce the tip speed of the rotor. Reductions in tip speed will reduce the main rotor noise and can be particularly important if the main rotor tip speed is already approaching transonic Mach numbers. The tip speed has an effect on all sources of rotorcraft noise. As a general design rule, the monopole source scales with the square of the tip speed, the dipole scales with the cube, and the quadrupole with the fourth power of tip speed [33, 34]. Figure 6.18a shows the effect of reducing the tip speed from a baseline value of $M_T = 0.6$ to $M_T = 0.4$ on the noise of a rotorcraft in climbing flight. Figure 6.18a shows the effect of reduced tip speed when the reduction in tip speed is from acoustic criteria only and when the rotor is re-trimmed at the same thrust at the lower tip speed. It is clear that the tip speed does significantly reduce the noise. Reductions are not as significant when we include the aerodynamic effects, as the rotor has to increase the sectional blade loading (through increases in blade pitch) to achieve the same thrust at the reduced tip speed.

Whilst the tip speed can reduce the noise, as we have seen, we must also balance the resulting effects on the aerodynamic performance. Typically, we compensate for reductions in tip speed with larger blade areas. However, this results in a larger rotor mass. Therefore, there are limitations to the reductions in tip speed before the increase in blade area becomes prohibitively heavy.

Rotor noise is also proportional to the blade count, with increasing blade counts reducing the acoustic emissions for the rotor. This reduction results from both acoustic criteria and the fact that the loading per blade is reduced. Figure 6.18b shows the resulting noise when the blade count is increased from 4 to 5 from both acoustic criteria and when the rotor is trimmed to the same thrust as the baseline configuration. We see that the increased blade count results in noise reductions from both acoustic and aerodynamic criteria. Further, the increased blade count can be beneficial for reducing the BVI source. With the increased blade count delivering the same thrust, the resulting tip vortex strength will be reduced and as will its subsequent interaction with the following blade resulting in a reduced impulse on the blade surface.

As with the tip speed, we must again consider the consequential effects of increasing the blade count, primarily an increase in rotor mass. Furthermore, increasing the blade count increases the blade passage frequency and will result in noise emissions that will be subjectively more annoying when corrected for frequency.

When reducing the tip speed is not possible, sweeping the blade tip may be useful in reducing the noise. The effect of sweep on the noise is twofold. Firstly, it has an effect similar to that of fixed-wing aircraft in that the relative velocity seen by the blade element is reduced proportionally to the sweep angle. Secondly, sweeping has a de-phasing effect. Where spanwise source emissions differ due to the chordwise displacement of the swept blade elements (see Ref. [35]).

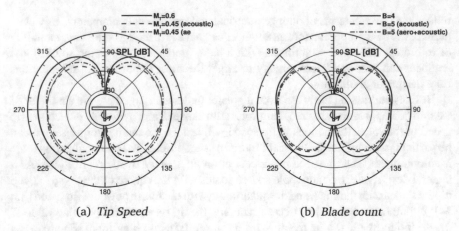

(a) *Tip Speed*   (b) *Blade count*

Fig. 6.18 Effects of design changes on resulting noise

The blade shape more generally, including the blade profile, sweep and dihedral or lean has played an important role in more modern rotorcraft designs, where careful optimisation of the tip shape has been utilised to reduce BVI noise. Brocklehurst and Barakos [10] present a review on the importance of rotorcraft tip shape design. The BlueEdge® blade is an excellent example of innovation that used unique tip shapes to reduce BVI noise and has been utilised on a number of Airbus helicopters, including the H160 [36–39].

It is possible to reduce rotor noise by encasing the rotor within a duct. Whilst the resulting excessive weight may be prohibitive for the main rotor, it has been success- fully utilised for the tail rotor. One further advantage of the duct is the ability to place acoustic liners within the duct to further reduce the noise at dominant frequencies. Whilst the duct may be promising, it must be weighed against the additional mass of the duct. A further disadvantage of the ducted rotor configuration is the need for struts to connect the tail rotor hub with the main tail duct. Whilst some of the disad- vantages of this can be alleviated by treating these struts as stator guide vanes, they will nonetheless give rise to unsteady interaction with the rotating tail rotor.

The Fenestron® is an example of a widely applied ducted tail rotor configuration shown to reduce noise. The Fenestron has a further innovation, in that the blades are non-equally spaced around the azimuth. This has the effect of distributing the blade tones over a greater frequency range and thus reducing the acoustic intensity. The same approach has been shown effective in propeller design [40, 41]. The applica- tion of non-equal spacing to the main rotor is somewhat more difficult due to the complexities of the main rotor, for example, the increased mass and that the blades are non-rigid.

Alongside the actual geometrical design of the rotorcraft, noise reductions can be realised by evaluating the operation of the rotorcraft. For example, BVI occurs at very specific decent glide slopes. If it is safe to increase the descent glide slope, BVI can be reduced. With increasing computational power, it is now viable to evaluate rotorcraft noise in manoeuvring flight [23, 25, 42–44]. This capability means that rotorcraft noise can be calculated over typical rotorcraft manoeuvres to more effectively evaluate the rotorcraft's community impact.

We have discussed a number of approaches and innovations that have been successfully applied to reduce rotorcraft noise. However, moving forward there will be new challenges that arrive in the world of rotorcraft acoustics. This is particularly true as we see the maturation of PAV or UAM technology. Filippone and Barakos present a good description of these challenges [45]. However, one particular challenge that arises with the increase in UAM technology is an increase in the design space. Whilst rotorcraft design space has changed little over its existence, UAM offer designer swathes of choice in the design of new novel rotorcraft. The designer now has a choice in rotor placement, spacing between rotors and the number of rotors, etc. and this presents a real challenge for the aeroacoustic engineer. This challenge mainly arises due to the fact that these new machines, if they are to be utilised within the urban environment, will have to meet very strict noise regulations. Furthermore, their commercial success, as we discussed in the introduction, may rely on the public acceptance of this new technology. As acoustic engineers, we must ensure that our tools are up to the task of modelling these new configurations, and particular challenges arise in the modelling of multiple rotors and their interaction with one another. With these new configurations, it is likely that the dominant acoustic sources will differ from that of the conventional rotorcraft and again we must ensure that we can correctly model these. In summary, these new architectures present a challenge for the acoustic engineer and will be the focus of much acoustic research as the technology matures.

In this chapter we have discussed the acoustics of rotorcraft. We started by looking at the reasons why rotorcraft noise is so important and how regulators enforce noise limits on new rotorcraft. We then looked at the sources of rotorcraft noise and how one may approach modelling some of these sources. We finally looked at some of the design choices for reducing rotorcraft noise. It is hoped that the reader has gained an appreciation of rotorcraft noise and is now equipped to tackle some preliminary evaluation and can build on this to tackle their own challenges in rotorcraft noise.

References

1. Ollerhead JB (1993) Past and present U.K. research on aircraft noise effects. In: National conference on noise control engineering, pp 9–18
2. British Standards Institution (2013) BS EN 61672-1:2013
3. Smith MJT (2004) Aircraft noise. Cambridge University Press
4. Filippone A (2012) Aircraft noise: noise sources. In: Advanced aircraft flight performance. Cambridge University Press, pp 470–532
5. Cox CR (1973) Aerodynamic sources of rotor noise. J Am Helicopter Soc 18(1):3–9
6. Schmitz FH (1991) Rotor noise. In: Aeroacoustics of flight vehicles, NASA TR 90-3052, pp 65–150
7. Brentner KS, Farassat F (1994) Helicopter noise prediction: the current status and future direction. J Sound Vib 170(1):79–96

8. Brentner KS, Farassat F (2003) Modeling aerodynamically generated sound of helicopter rotors. Prog Aerosp Sci 39(2–3):83–120
9. Filippone A, Afgan I (2008) Orthogonal blade-vortex interaction on a helicopter tail rotor. AIAA J 46(6):1476–1489
10. Brocklehurst A, Barakos GN (2013) A review of helicopter rotor blade tip shapes. Prog Aerosp Sci 56:35–74
11. Lowson MV (1996) Focusing of helicopter BVI noise. J Sound Vibr 190(3):477–494
12. Yu YH (2000) Rotor blade-vortex interaction noise. Prog Aerosp Sci 36(2):97–115
13. Gutin L (19848) On the sound field of a rotating propeller. NACA TM 1195
14. Ffowcs Williams JE, Hawkings DL (1969) Sound generation by turbulence and surfaces in arbitrary motion. Philos Trans R Soc A Math Phys Eng Sci 264(1151):321–342
15. Lighthill MJ (1952) On sound generated aerodynamically. Proc R Soc Lond A 211:564–587
16. Farassat F (1994) Introduction to generalized functions with applications in aerodynamics and aeroacoustics. NASA Technical paper 3428
17. Brentner KS, Farassat F (1998) Analytical comparison of the acoustic analogy and Kirchhoff formulation for moving surfaces. AIAA J 36(8):1379–1386
18. Brentner KS (1988) Prediction of helicopter rotor discrete frequency noise for three scale models. J Aircr 25(5)
19. Brentner KS, Burley CL, Marcolini MA (1994) Sensitivity of acoustic predictions to variation of input parameters. J Am Helicopter Soc 39(3):43–52
20. Farassat F (1975) Theory of noise generation from moving bodies with and application to helicopter rotors. NASA TR R-451
21. Farassat F (1981) Linear acoustic formulas for calculation of rotating blade noise. AIAA J 19:1122–1130
22. Farassat F, Succi GP (1980) A review of propeller discrete frequency noise prediction technology with emphasis on two current methods for time domain calculations. J Sound Vibr 71(3):399–419
23. Brentner KS, Jones HE (2000) Noise prediction for maneuvering rotorcraft. In: 6th AIAA/CEAS Aeroacoustics conference
24. Lopes LV (2015) Compact assumption applied to the monopole term of Farassat's formulations. In: 21st AIAA/CEAS Aeroacoustics conference
25. Brentner KS, Bres GA, Perez G, Jones HE (2002) Maneuvering rotorcraft noise prediction? A new code for a new problem. In: AHS Aerodynamics, acoustics, and test evaluation specialist meeting, San Francisco, CA
26. Brentner KS (1996) Numerical algorithms for acoustic integrals—the devil is in the details. In: 2nd AIAA/CEAS Aeroacoustics conference
27. Casper J, Farassat F (2002) A new time domain formulation for broadband noise predictions. Int J Aeroacoustics 1(3):207–240
28. Brooks TF, Burley CL (2001) Rotor broadband noise prediction with comparison to model data. In: 7th AIAA/CEAS Aeroacoustics conference and exhibit
29. Amiet RK (1975) Acoustic radiation from an airfoil in a turbulent stream. J Sound Vibr 41(4):407–420
30. Amiet RK (1986) Airfoil gust response and the sound produced by airfoil-vortex interaction. J Sound Vibr 107(3):487–506
31. Pegg RJ (1979) A summary and evaluation of semi-empirical methods for the prediction of helicopter rotor noise. NASA TM 80200
32. Brooks TF, Pope DS, Marcolini MA (1989) Airfoil self-noise and prediction. NASA Reference publication 1218
33. Lowson MV (1966) Basic mechanisms of noise generation by helicopters, VSTOL aircraft and ground effect machines. J Sound Vibr 3(3):454–466
34. Brown D, Ollerhead JB (1971) Propeller noise at low tip speeds. Wyle Laboratories, WR-71-9
35. Hanson DB (1980) Influence of propeller design parameters on far-field harmonic noise in forward flight. AIAA J 18(11):1313–1319

36. Rauch P, Gervais M, Cranga P, Baud A, Hirsch JF, Walter A, Beaumier P (2011) Blue edge™: the design, development and testing of a new blade concept. In: 67th Annual forum of the American helicopter society, pp 542–555
37. Bebesel M, D'Alascio A, Schneider S, Guenther S, Vogel F (2015) Bluecopter demonstrator—an approach to eco-efficient helicopter design. In: 41st European rotorcraft forum, Munich, Germany
38. Schneider S, Heger R, Konstanzer P (2016) BLUECOPTER demonstrator: the state-of-the-art in low noise design. In: 42nd European rotorcraft forum, Lille, France
39. Guntzer F, Gareton V, Pinacho JP, Caillet J (2019) Low noise design and acoustic testing of the airbus helicopter H160-B. In: 45th European rotorcraft forum, Warsaw, Poland
40. Barakos GN, Johnson CS (2016) Acoustic comparison of propellers. Int J Aeroacoustics 15(6–7):575–594
41. Chirico G, Barakos GN, Bown N (2018) Numerical aeroacoustic analysis of propeller designs. Aeronautical J 122(1248):283–315
42. Chen H, Brentner K, Lopes L, Horn J (2004) A study of rotorcraft noise prediction in maneuvering flight. In: 42nd AIAA Aerospace sciences meeting and exhibit, Reno, Nevada
43. Vouros S, Goulos I, Pachidis V (2019) Integrated methodology for the prediction of helicopter rotor noise at mission level. Aerosp Sci Technol 89:136–149
44. Greenwood E, Rau R (2020) A maneuvering flight noise model for helicopter mission planning. J Am Helicopter Soc 65(2)
45. Filippone A, Barakos GN (2021) Rotorcraft systems for urban air mobility: a reality check. Aeronautical J 125(1283):3–21

Dr. Dale Smith, having pursued a Ph.D. at the University of Manchester, is now working at Qinetiq. He specialises in propeller and counter-propeller aerodynamics and aero-acoustics.

Prof. George Barakos is at the School of Engineering, Glasgow University, where he specialises in high-fidelity aerodynamics and aero-acoustics models, computational fluid dynamics, high-performance computing and rotorcraft engineering.

Chapter 7
Rotorcraft Control Systems

Rafael Morales

Abstract This chapter provides an overview of flight control systems for rotorcraft at an introductory level. There exist many rotorcraft configurations, however, we will focus our attention on conventional helicopters to provide a fundamental understanding. We will cover standard and advanced control design methods to design and validate flight control algorithms for both stability augmentation and autopilot systems. We will touch on some essential elements of feedback control theory before showing how the design methods are implemented. A key characteristic of the discussed design methods is that they are suitable for multivariable systems, which offer advantages for minimising key helicopter dynamic couplings. The control design methods also offer improved robustness properties leading to flight envelope protection characteristics. We will cover key metrics to assess the robustness and performance properties of the flight control laws from a control theory approach. These control laws form the basis for the assessment of the flight control laws in terms of handling qualities.

Nomenclature

BIBO	Bounded-Input Bounded-Output
LQR	Linear Quadratic Regulator
LTI	Linear Time-Invariant
MIMO	Multiple-Input Multiple-Output
PID	Proportional-Integral-Derivative
SISO	Single-Input Single-Output
A, B, C, D	Matrix coefficients of state-space representations
F	Output feedback gain matrix

Supplementary Information The online version contains supplementary material available at https://doi.org/10.1007/978-3-031-12437-2_7.

R. Morales (✉)
School of Engineering, University of Leicester, Leicester, UK
e-mail: rm23@leicester.ac.uk

H	State feedback gain matrix
I	Identity matrix with adequate dimensions
G	Plant
K	Compensator
$G(s)$	Plant transfer function
$K(s)$	Controller transfer function
$S(s)$	Sensitivity transfer function
$T(s)$	Co-sensitivity transfer function
$w_p(s)$	Sensitivity weight
$w_{KS}(s)$	Control actions weight
$w_T(s)$	Co-sensitivity weight
k_p, k_i, k_D, N	PID controller parameters
S_o^*, M, ω_S^*	Sensitivity weight parameters
$\omega_T^*, N_T, \gamma_T$	Co-sensitivity weight parameters
w_{KS}	Control actions constant parameter
j	Imaginary unit
l	Number of outputs
m	Number of inputs
n	Number of states
s	Laplace variable
t	Time (s)
T	Transpose operator
$c(t)$	Command or reference signal
$e(t)$	Error signal
$n(t)$	Noise signal
$u(t)$	Control actions
$x(t)$	State vector signal
$y(t)$	Output signal
$\bar{\sigma}(.)$	Largest singular value
$\underline{\sigma}(.)$	Lowest singular value
ω	frequency variable (rad/s)
ω_S	Sensitivity bandwidth (rad/s)
ω_T	Co-sensitivity bandwidth (rad/s)
$\|.\|_\infty$	Infinity norm

7.1   Feedback Control

In the context of engineering systems, the fundamental objective of feedback control is to regulate flight variables of interest and de-sensitise the rotorcraft dynamics to key parameter variations (e.g. changes in centre of gravity, aerodynamic coefficients, etc.). Regulation of key aircraft flight variables (attitude and velocities) are related to pilot command tracking and alleviation of gust perturbations as well. There are three main characteristics when looking at the design of flight control systems:

- **Stabilisation**: This property is concerned with the ability to stabilise the dynamics of the rotorcraft. A primary role of using feedback in flight control systems is to stabilise rotorcraft flight dynamics and/or improve the original rotorcraft transient response. This is one of the main roles of the subsystem known as stability augmentation.
- **Performance**: The performance of the flight control systems is measured with respect to the ability to track pilot command signals and provide disturbance rejection capabilities. These two properties can be assessed separately. There are a number of key metrics developed to assess the quality of the performance, both in the time- and the frequency-domain. This is one of the main roles of the flight control subsystem known as autopilot or autonomous system.
- **Robustness**: This property is concerned with the ability of the control algorithm to maintain stability and certain level of performance despite variations on the rotorcraft aeromechanics. In control theory, this property is further split into robust stability and robust performance but we will consider only robust stability in this chapter to simplify the discussions [1]. Although robust stability might not get much attention in flight dynamics textbooks, robust stability is of paramount importance to flight control because of its impact on the airworthiness and certification requirements of the aircraft.

The afore-mentioned characteristics provide great benefits in terms of pilot work load alleviation, which in turn also has an impact on improving the safety of operations. Given significant progress in the areas of Artificial Intelligence and Machine Learning methods [2], there has been an increased interest recently to apply these methods to replace key pilot tasks, such as obstacle avoidance and trajectory planning. In the aerospace community these properties are referred as autonomy. Autonomy algorithms are outside the scope of this chapter but note that their performance is strongly dependent on the performance and reliability of the flight control systems we discuss in this chapter.

The afore-mentioned qualities are achieved in flight control systems using feedback loops, with the elementary representation shown in Fig. 7.1. This feedback loop shows the special interconnection of two systems: G represents the system to control and the controller K. The flight variables of interest are collected in the vector signal $y(t)$. Measurements of the output signal is represented by $y_m(t)$ after the presence of measurement noise $n(t)$ has been included. The difference between commands $c(t)$ and the measured control signal $y_m(t)$ is then fed into a compensator K or algorithm to process or dictate the various control signals represented by $u(t)$ to perform the control task. In this diagram note that the signal $d(t)$ accounts for external disturbances, which for autopilot applications, accounts for gusts disturbances typically.

Flight control systems exploit the benefits of feedback and use a nested configuration for stability augmentation and pilot command tracking tasks, see Fig. 7.2. The inner feedback loop, known as stability augmentation, is included to provide

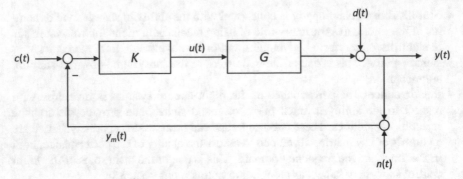

Fig. 7.1 Feedback interconnection

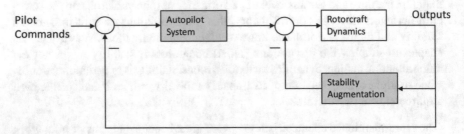

Fig. 7.2 Standard architecture for flight control systems

improved flight dynamics. These are required for instance if the lateral and longitu-dinal dynamics of the original aircraft are either unstable or provide poor transient response (low-damped modes and/or slow responses). Improving the flight dynamics of the rotorcraft facilitates the design and enables achieving a desired performance for pilot command tracking purposes. This is the main purpose of this outer loop con-trol system, also known as the autopilot system. The performance is assessed mainly in terms of tracking and disturbance rejection capabilities, which are connected to handling quality requirements.

7.2 Flight Dynamic Models for Rotorcraft Flight Control Design

Rotorcraft flight dynamics obtained from first principles are complex (high-order and nonlinear) and in most cases not suitable for standard control system design methods. For this reason, available models, which are usually obtained following first principle models [3], need to be approximated to a linear form to facilitate the design of the flight control algorithms. These linear representations are only valid for very spe-cific points of the flight envelope and obtained at trimmed (equilibrium) conditions. Linear models hence provide a *local* description of the aircraft behaviour around a chosen equilibrium position. For instance, for conventional helicopter, the main flight conditions are hover, vertical motion, longitudinal and lateral flight, trimmed at different airspeeds. Multiple linearisation is required to obtain a simplified and comprehensive, yet meaningful, picture of the rotorcraft behaviour.

Simplified rotorcraft flight dynamics models are expressed as a Linear-Time-Invariant (LTI) system. LTI systems can be represented as either a transfer function matrix, typically denoted as $G(s)$, or alternatively, as a state-space representation

$$G \sim \begin{cases} x(t) = Ax(t) + Bu(t), & x(0) \in \mathbb{R}^n \\ y(t) = Cx(t) + Du(t) \end{cases} \tag{7.1}$$

The coefficient n denotes the number of states required to represent the rotorcraft flight dynamics and represents also the dimension of the vector signal (in column form) $x(t)$. The number of available measured outputs is l and represents also the dimension of the output signal $y(t)$. The number of control inputs to operate the rotor-craft is m which is also the dimension of the vector signal $u(t)$. The matrix coefficients of the model are constant and their dimensions are $A \in \mathbb{R}^{n \times n}$, $B \in \mathbb{R}^{n \times m}$, $C \in \mathbb{R}^{l \times n}$ and $D \in \mathbb{R}^{l \times m}$. Typically, $y(t)$ contains a subset of the signals in $x(t)$ and hence $l < n$ and $D = 0$.

A very important relation that connects the transfer function and state-variable representations is

$$G(s) = C(sI - A)^{-1}B + D \tag{7.2}$$

A key concept in control theory on whether a state-space representation is totally equivalent to the transfer function matrix representation shown above is associated with more fundamental concepts known as *controllability* and *observability*. To simplify the discussions in this chapter, we will assume that any obtained flight dynamic model represented in state-space form is both controllable and observable. This assumption also applies for the representation of the controller K. For more information on these concepts, refer to [1].

Another important concept associated with LTI systems is stability. Stability of LTI systems can be considered from two points of views - the quality of the free and forced response. Stability of the free response (zero inputs) is considered for non-zero initial conditions and analysed via state-space representations typically. Loosely speaking, we say the free response is stable if the state vector (and hence the output) remain finite at all times and converge to zero. This stability notion is known as asymptotic stability. On the other hand, stability of the forced response (zero initial conditions) is considered via transfer function representations. Loosely speaking, we say that the forced response is stable if finite input signals leads to finite output signals. This stability concept is referred to as Bounded-Input Bounded-Output (BIBO) stability. Asymptotic stability is obtained if the eigenvalues of the matrix A have negative real part. For transfer functions, BIBO stability is obtained if all the roots of the denominator of the matrix $G(s)$ have negative real part. For LTI representations which are controllable and observable, the roots of the denominator of $G(s)$ are the same as the eigenvalues of the state matrix A. Therefore we can claim

Table 7.1 Typical state-vector and control elements in conventional helicopters

State	Description
$\theta(t)$	Pitch attitude
$\phi(t)$	Roll attitude
$p(t)$	Pitch rate
$q(t)$	Roll rate
$r(t)$	Yaw rate
$v_x(t)$	Longitudinal or forward velocity component
$v_y(t)$	Lateral velocity component
$v_z(t)$	Vertical velocity component

Table 7.2 Control inputs in conventional helicopters

Control input	Description
$u_{col}(t)$	Main rotor collective
$u_{long}(t)$	Longitudinal cyclic
$u_{lat}(t)$	Lateral cyclic
$u_{tail}(t)$	Tail-rotor collective

that these two stability notions are equivalent in the sense that the same condition must hold to ensure both types of stability. These stability conditions are necessary and sufficient meaning that if they are not satisfied, then the LTI system do not meet the definitions of asymptotic and BIBO stability [4].

Typically for conventional helicopters, the state vector contains the signals shown in Table 7.1 and the control inputs are shown in Table 7.2. Note that the adopted notation for the linear velocities is different from the more standard notation found on flight dynamics textbooks to avoid confusion with the notation implemented to represent signals in the feedback loop.

Rotorcraft flight dynamics are usually separated into longitudinal and lateral dynamics when performing dynamic stability analysis and flight control design [3, 5]. Longitudinal dynamics consider the pitch and vertical motion of the rotorcraft, while lateral dynamics are concerned with the yaw or directional motion and the rolling behaviour. This separation might not be valid for all rotorcraft configurations and hence a more comprehensive multivariable model that models cross-couplings between these two dynamics would be required to design a more effective flight control system at minimising undesired couplings.

7.3 Stability, Robustness and Performance of Feedback Control Systems

7.3.1 Feedback Stability

The primary property to assess in any feedback control system is its stability. To simplify the discussions, we will assume that both the compensator K and the plant G can be represented by LTI elements reliably. Feedback stability refers to the values of the signal at any location in the feedback loop remaining finite in the presence of finite exogenous signals $(c(t), d(t), n(t))$. For this reason it is necessary to look at various transfer functions obtained from all possible pairs of inputs and outputs in the feedback loop. A sufficient and necessary condition for stability of the feedback loop is simplified by testing for stability of the following transfer functions

$$(I + K(s)G(s))^{-1} \tag{7.3}$$
$$K(s)(I + G(s)K(s))^{-1} \tag{7.4}$$
$$G(s)(I + K(s)G(s))^{-1} \tag{7.5}$$
$$(I + G(s)K(s))^{-1} \tag{7.6}$$

Refer to [1] for more details. A more elaborate and general condition for feedback stability can also be found in [6], which is expressed in terms of state-space representations for G and K. Feedback systems with this property are called *internally stable* and this stability test should be verified before performing any performance assessment of the flight control system.

7.3.2 Feedback Robustness

Stability alone is not enough in rotorcraft control system design mainly because LTI models can only capture the dynamics of the rotorcraft up to some extent and rotorcraft dynamics are also subject to variations. For this reason, we are interested in determining somehow how tolerant the feedback loop is to certain variations in the flight control system. There are two main metrics that are widely used for this purpose - they are the Gain Margin (GM) and Phase Margin (PM). These metrics are derived from the well-known Nyquist Stability Criterion [4] but these metrics are applicable only to single-input single-output feedback systems, i.e., when both G and K have one input and one output and both are stable in most cases.

GM accounts for gain variations in G typically and this metric indicates how much the gain can be increased before the feedback loop become unstable. A typical design requirement for feedback systems is

$$GM \geq 2 \tag{7.7}$$

The PM is associated instead with time delays expressed in the form of phase-lag. Delays are expected to occur in the communication channel between the controller and any actuators primarily and latency from on-board sensors. The PM indicates how much phase lag can exist before the system becomes unstable. A typical design requirement is

$$PM \geq 30° \tag{7.8}$$

GM and PM are not always reliable since they account for individual changes in gain and delays, but not combined or simultaneous variations. For this reason a less known and very useful criterion is provided in terms of a metric known as the peak sensitivity. The Sensitivity transfer function $S(s)$ of the feedback loop is defined as

$$S(s) = (I + G(s)K(s))^{-1} \tag{7.9}$$

Loosely speaking, the largest gain of this sensitivity when evaluated in the frequency domain is indicated by

$$\|S\|_\infty = \max_\omega |S(j\omega)|, \forall\omega \tag{7.10}$$

This value is also known as the H-infinity norm of the Sensitivity. The reasons for this particular notation is outside the scope of this manuscript, however, for those interested Readers, they can refer to the mathematical field of Functional Analysis [7]. A typical robustness condition in terms of the peak sensitivity is

$$\|S\|_\infty \leq 2 \tag{7.11}$$

For multivariable systems, a metric for robustness can be obtained from the multivariable Nyquist criterion. Note that the stability condition of the closed-loop can be expressed in the frequency domain as follows

$$\det(I + G(j\omega)K(j\omega)) \neq 0, \forall\omega \tag{7.12}$$

and for stable $G(s)$ and $K(s)$. This means that the plot of $\det(I + G(j\omega)K(j\omega))$ must not enclose the origin to ensure stability. The robustness metric can then be obtained by making sure the plot do not pass too close to the origin of the complex plane. A similar metric derived from the peak sensitivity mentioned above could be established by making sure the closest point to the origin is larger than 0.5

$$|\det(I + G(j\omega)K(j\omega))| \geq 0.5, \forall\omega \tag{7.13}$$

Clearly the more the above conditions are exceeded, the more robust the flight control system is. However, the conflicting nature between performance and robustness is well-know in feedback systems and the task of the control design engineer is to achieve a desired trade-off. For practical flight control systems where the feedback loops are SISO, it is recommended to use the three aforementioned metrics

(GM, PM and $\|S\|_\infty$) whenever applicable. The metrics can be easily calculated with commercial software such as Matlab.

7.3.3 Performance Assessment: Time-domain

A pragmatic approach to evaluate the performance of the control system system is by simulating the responses to step command and disturbance signals and in the presence of white noise. This approach is particularly important to assess the transient characteristics of the flight control system. For LTI feedback systems, the analysis can be done separately due to the superposition principle that governs LTI systems but this is not the case if the controller is implemented on more comprehensive nonlinear models. When considering step responses, there are well known metrics, such as overshoot, rise time, steady-state error and settling time, that should be used when assessing such responses [4]. These metrics are introduced usually to inspect the tracking characteristics of control systems in most textbooks but they can be applied also to assess the disturbance rejection characteristics of the control system [4].

Apart from step signals, the control system should be tested against all possible signal forms according to the considered application. The performance assessment should also include the effects of measurement noise which is usually simulated by using white noise signals in $n(t)$. Also, any time-domain simulations should inspect the control signal $u(t)$ to make sure it complies with actuators operational capabilities. All these aspects are very important and they all should be tested in any simulation campaign very carefully and extensively. The more comprehensive the assessment of the control system is under simulation environments, the more confidence is gained before the system is tried on practical implementations.

7.3.4 Performance Assessment: Frequency-Domain

The performance of feedback control systems should also be assessed in the frequency domain because the information is richer and more comprehensive than in the time-domain. Assessment in the frequency domain is particularly important to assess the performance at steady-state. For this purpose, note that the feedback loop depicted in Fig. 7.1 can be described by the following relations

$$y(s) = T(s)c(s) + S(s)d(s) + T(s)n(s) \qquad (7.14)$$
$$e(s) = c(s) - y(s) = -S(s)c(s) + S(s)d(s) - T(s)n(s) \qquad (7.15)$$
$$u(s) = K(s)S(s)(c(s) - d(s) - n(s)) \qquad (7.16)$$

Note that the above expressions are expressed in terms of two especial transfer functions: (i) the Sensitivity $S(s)$ defined in (7.9) and (ii) the Complementary Sensitivity or co-sensitivity $T(s)$ defined as follows

$$T(s) = G(s)K(s)(I + G(s)K(s))^{-1} \qquad (7.17)$$

The response of the output to command signals and noise is determined by $T(s)$ and hence this is the transfer function to assess for pilot command tracking tasks if perturbations are not significant. On the other hand, the effect of external disturbances on the output and the error signal is captured by the Sensitivity transfer function $S(s)$. The response of the control actions $u(t)$ due to the presence command, perturbations and noise are dictated by $K(s)S(s)$. Therefore these three transfer functions play a key a role in the performance of the flight control system and all of them should be assessed very carefully when validating any control strategy.

The more general way to assess the performance is by inspecting the singular values [8] of the transfer functions in the frequency domain. $\sigma(S(j\omega))$ denotes the singular value of the complex matrix $S(j\omega)$. A similar notation follows for $\sigma(T(j\omega))$ and $\sigma(K(j\omega)S(j\omega))$. The smallest and largest singular values are denoted by $\underline{\sigma}(.)$ and $\overline{\sigma}(.)$, respectively. The singular value plot for a transfer function matrix plots the singular values across all frequencies. This plot can be obtained with the commercial software Matlab using the command $\texttt{sigma}$. There are key parameters flight control design engineers should take into account when assessing the performance in the frequency domain. We bring our attention to the following metrics:

Sensitivity bandwidth. This parameter is denoted by ω_S and it is expressed in rad/s usually. This metric is obtained by the frequency at which the largest singular value of $S(j\omega)$ crosses -3 dB from below

$$\overline{\sigma}(S(j\omega_S)) = \frac{1}{\sqrt{2}} \approx 0.707$$

This metric can be considered as the closed-loop bandwidth, and it provides a metric of the largest frequency component in the disturbance signal $d(t)$ for which satisfactory attenuation is achieved. This is also the largest frequency at which the error signal $c(t) - y(t)$ is reduced satisfactorily. It is therefore expected that for any disturbance containing frequency elements above this value, the disturbance rejection and the error between the command signal and the controlled variable become poor.

Peak Sensitivity. Recall that this metric $\|S\|_\infty$ was already introduced earlier as a robustness parameter for SISO systems. For MIMO system, the peak sensitivity becomes the maximum of the largest singular value of the Sensitivity across all frequencies

$$\|S\|_\infty = \max_\omega \overline{\sigma}(S(j\omega)) \qquad (7.18)$$

This metric represents a measurement of the worst-case gain when performing disturbance rejection. Clearly, we would like to have this value as low as possible.

Sensitivity Low-frequency Gain. This metric represents the worst gain expected when providing disturbance rejection to constant signals and we will denote it as

$$S_0 = \overline{\sigma}(S(j0)) \tag{7.19}$$

In flight control examples, this metric would be related to the the capabilities of the flight control system to reject constant gusts. We would like to design the flight control system to achieve as low a value as possible of S_0.

Co-sensitivity bandwidth. This parameter is denoted by ω_T and is defined when the lowest singular value of $T(j\omega)$ crosses -3 dB from above

$$\underline{\sigma}(T(j\omega_T)) = \frac{1}{\sqrt{2}} \approx 0.707 \tag{7.20}$$

This metric indicates the maximum frequency component in the command signals for which satisfactory tracking is achieved. Note that this metric is only concerned with tracking of the amplitude for command signals, but does not incorporate information about the quality of phase tracking. For this reason this definition of bandwidth can be misleading. For example, consider the SISO case where the singular value becomes $|T(j\omega)|$. We can have a metric of bandwidth obtained at $|T(j\omega_T)| \approx -3$ dB, but on the other hand having a poor phase tracking (for instance $\angle T(j\omega_T) \geq 30°$). In this case the tracking of the harmonic command signal would be poor because of the noticeable phase difference between the output and command signals, and a more reliable bandwidth metric would have a much lower value than the bandwidth measurement provided by the above definition.

7.4 Control Design for Flight Control

There exist a plethora of control design methods with particular benefits depending on the requirements of the application. We will discuss in this section a selected combination of conventional and more recent design methods given the benefits they bring to the design of both stability augmentation and autopilot systems in flight control system design.

7.4.1 Stability Augmentation

The main purpose of control design for stability augmentation is the improvement of the rotorcraft dynamics by the use of feedback. Rotorcraft dynamic are adjusted depending on the desired degree of rotorcraft stability. A common approach is to use feedback to relocate key rotorcraft modes into desired locations, thus improving their transient characteristics by typically making them more stable, (negative real part being larger in absolute value), reducing the damping (reducing their imaginary part or making their locations closer to the real axis) and stabilising unstable modes (moving these poles into the left-half side of the complex plane). Two common strategies for these two goals are Root Locus and Pole Placement [4]. While Root Locus is particularly useful for SISO control system design, Pole Placement is more general and suitable for MIMO systems.

Stability augmentation can be achieved typically by feeding the attitude rates and using these measurements for pole placement. The method relies on manipulation of the state-space description and makes full use of appropriate computational tools. Its real application is limited by the assumption that reliable measurements or estimations of the state-variables are available. The *unaugmented* or open-loop rotorcraft dynamics are expressed in the standard state-representation (7.1). The pole relocation is implemented by means of a full-state feedback law expressed in the form of a linear combination of the states

$$u(t) = v(t) - Hx(t) \tag{7.21}$$

The feedback gain matrix $H \in \mathbb{R}^{m \times n}$ determines the location of the closed-loop and is the parameter to choose. The new input signal $v(t)$ is the input to the augmented system and is the new input used later on for the autopilot design task. The above control law leads to the following closed-loop or augmented state-space representation

$$\dot{x}(t) = (A - BH)x(t) + Bv(t) \tag{7.22}$$
$$y(t) = (C - DH)x(t) + Dv(t) \tag{7.23}$$

There exist computational tools such as Matlab (see command `place`) whereby the user provides the values of A, B and the desired closed-loop pole locations in vector form so the eigenvalues of the closed-loop matrix $A - BH$ match the desired modes. The command will return the numerical values of H. There are certain requirements on A and B for the strategy to work, such as the pair A and B being a controllable pair, the number of closed-loop poles must be the same as n and no re-located eigenvalue should have a multiplicity greater than the number of inputs [1].

In many rotorcraft applications the output signal $y(t)$ is a subset of the elements in $x(t)$, leading to a zero matrix D. The pole placement approach can be implemented via Output feedback in these cases in a simple form. Note the control law becomes instead

$$u(t) = v(t) - Fy(t) \tag{7.24}$$

leading to the following closed-loop description

$$\dot{x}(t) = (A - BH)x(t) + Bv(t) \tag{7.25}$$
$$y(t) = Cx(t) \tag{7.26}$$

with

$$H = FC \tag{7.27}$$

The pole placement can be implemented by using the original matrices A, B and specifying the desired closed-loop locations to obtain H as explained earlier. The gain matrix F can be obtained as follows

$$F = HC^\dagger \tag{7.28}$$

whereby $C^\dagger = C^T(CC^T)^{-1}$ represents the pseudo-inverse of C.

7.4.2 Autopilot System

Recall the outer-loop of the flight control system provides autopilot characteristics enabling the pilot to execute certain manoeuvres under automatic control. The autopilot system controls the rotorcraft motion via the regulation of the rotorcraft attitude. Even basic automation capabilities might look limited compared with more advanced autonomy algorithms, yet autopilot systems reduce pilot workload significantly. A common approach to tune SISO control systems is Proportional-Integral-Derivative (PID) control for the tracking of the attitude. The control law following PID principles is implemented in practice typically as

$$K(s) = k_p + \frac{k_I}{s} + \frac{k_d s}{Ns + 1} \tag{7.29}$$

The controller parameters are k_p, k_i, k_d and N. The controller in this case will provide the actions in terms of the new control input $v(t)$ introduced by the stability augmentation system.

A design approach suitable to rotorcraft flight control when dynamic couplings are significant and can not be addressed by the stability augmentation system are based on state-space methods and the more recent Robust Control methods. A popular approach known as Linear Quadratic Regulator (LQR) could be implemented to handle the challenges caused by the multivariable control design problem. However, these controllers rely on accurate estimation or measurements of the states (not only the attitude angles) and these controllers are criticised for a lack of robustness [9]. We explore a more recent design strategy which are based on the principle of shaping key closed-loop transfer functions to achieve a desired level of robustness and performance. The particular design approach that we will consider is covered under the umbrella of Robust Control methods, which are also known as $\mathcal{H}_\infty$ control. The design methodologies rely on advanced optimisation algorithms to obtain the control laws but the design procedures of the controllers are rather transparent to the design engineer without the need to know the intricacies of the optimisation routines.

Mixed-Sensitivity Robust Control aims to shape the frequency response of three key transfer functions: the Sensivity $S(s)$, the Co-Sensitivity $T(s)$ and $K(s)S(s)$ which accounts for the control actions. The principles are based on using transfer function weights to dictate the desired shape.

Recall that the lower the values of $\overline{\sigma}(S(j\omega))$, the better characteristics in terms of disturbance rejection and error reduction. It is therefore natural to indicate the desired shape of the Sensitivity by setting an upper bound on desired vales across frequencies. Mathematically, we can express this as

$$\overline{\sigma}(S(j\omega)) < |w_p(j\omega)|^{-1}, \quad \forall\omega \tag{7.30}$$

The inverse on the transfer function on the right hand side is introduced for mathematical convenience, and $w_p(s)$ in this case is considered as a SISO transfer function to facilitate the discussions. The above inequality can be equivalently expressed in terms of a weighted Sensitivity

$$\|w_p S\|_\infty < 1 \tag{7.31}$$

A common choice for $w_p(s)$ provides large gain inside the desired control bandwidth to demand low sensitvity gains

$$w_p(s) = \frac{s/M + \omega_S^*}{s + S_0^* \omega_S^*} \tag{7.32}$$

The parameters could be chosen for instance to prescribe design requirements, such as

$$S_0^* \leq \text{Desired worst-case steady-state error}$$
$$M \leq \text{Desired worst-case disturbance rejection gain}$$
$$\omega_S^* \geq \text{Desired minimum bandwidth (rad/s)}$$

The condition to shape the frequency response for the control actions transfer function $K(s)S(s)$ is as follows

$$\|w_{KS} KS\|_\infty < 1 \tag{7.33}$$

In many cases, a simple constant weight can be chosen to determine the desired largest control actions given largest amplitude values of the exogenous signals ($c(t), d(t)$ and $n(t)$). For instance, in the SISO case and considering only the effects of command signals, the weight can be chosen as

$$\overline{u} \leq w_{KS}^{-1} \overline{c} \tag{7.34}$$

with $\bar{u}$ and $\bar{c}$ representing the desired largest absolute control input amplitude and the expected largest command amplitude value, respectively. Note that the above inequality holds only for steady-state values and hence the constraint on the control input at every time instant is not guaranteed but it can be used as an initial choosing value for w_{KS}.

The same principles can be applied to shape the co-sensitivity $T(s)$ via a performance weight $w_T(s)$

$$\|w_T T\|_\infty < 1 \tag{7.35}$$

Typically the weight offer large gain outside the bandwidth region to ensure good noise attenuation and also to provide good robustness characteristics [1]. A weight that meets these specifications has the following structure

$$w_T(s) = \gamma_T \frac{s + \omega_T^*}{s + N_T \omega_T^*} \tag{7.36}$$

with

$\omega_T^* \geq$ Maximum frequency at which dynamic model is considered reliable

$N_T > 1$

$\gamma_T \geq$ Amount of relative uncertainty at high frequencies

In many applications, the control design are based on shaping the frequency responses of $S(s)$ and $K(s)S(s)$ only, providing very good design results. The control Engineer can use computational tools such as those found in the Robust Control Toolbox in Matlab to be able to obtain the controllers which satisfy one or more of the above specification requirements. The Matlab command to perform mixed-sensitivity control design is `mixsyn`. There are many more mathematical details which are not discussed in this chapter around this control design philosophy but the Reader is referred to [1] for a comprehensive source on robust control methods. The main appeal of the above methods in comparison with more traditional approaches, such as PID, is that there is a more transparent relation between the controller parameters, performance and robustness specifications and a desired trade-off. We will demonstrate the aforementioned design strategies with a numerical example shown in the next section.

7.5 Rotorcraft Flight Control System Design—Numerical Example

For this example, we will use the helicopter model found in Matlab under the title "Multi-loop controller of a helicopter". This example uses an eight-state helicopter model at the hovering trim condition. The model is presented in state-space form, with the state vector having the same signals shown in Table 7.1 arranged as follows

$$x(t) = [v_x(t), v_z(t), q(t), \theta(t), v_y(t), p(t), \phi(t), r(t)]^T \tag{7.37}$$

In this example, the linear velocity are expressed in m/s. Attitude angles and rates are expressed in deg and deg/s, respectively. The control inputs in this example are the longitudinal and lateral cyclic, as well as the tail rotor

$$u(t) = [u_{long}(t), u_{lat}(t), u_{tail}(t)]^T \tag{7.38}$$

The command signals are expressed in degrees. The values of the state-space model are as follows

$$A = \begin{bmatrix} -0.0191 & 0.0170 & 0.3839 & -9.7924 & -0.0008 & -0.3371 & 0 & 0 \\ 0.0136 & -0.2994 & 0.0237 & -0.5859 & -0.0017 & -0.0257 & 0.5374 & 0 \\ 0.0405 & -0.0026 & -1.8394 & 0 & 0.0024 & 0.5281 & 0 & -0.0015 \\ 0 & 0 & 0.9985 & 0 & 0 & 0 & 0 & 0.0549 \\ 0.0010 & -0.0017 & -0.3381 & 0.0322 & -0.0349 & -0.4032 & 9.7777 & 0.1168 \\ 0.0130 & 0 & -3.047 & 0 & -0.229 & -10.6199 & 0 & -0.0333 \\ 0 & 0 & -0.0033 & 0 & 0 & 1 & 0 & 0.0598 \\ 0.0020 & 0.0060 & -0.5412 & 0 & 0.0039 & -1.8554 & 0 & -0.3487 \end{bmatrix}$$

$$B = \begin{bmatrix} -10.3456 & 1.0793 & 0 \\ -0.7293 & 0.0755 & 0 \\ 27.0900 & -4.7239 & -0.1857 \\ 0 & 0 & 0 \\ -1.0820 & -10.3713 & 4.7239 \\ -27.2884 & -156.4425 & -1.0690 \\ 0 & 0 & 0 \\ -4.8969 & -27.9728 & -12.9304 \end{bmatrix}, C = \begin{bmatrix} 0 & 0 & 0 & 1 & 0 & 0 & 0 & 0 \\ 0 & 0 & 0 & 0 & 0 & 0 & 1 & 0 \\ 0 & 0 & 0 & 0 & 0 & 0 & 0 & 1 \\ 0 & 1 & 0 & 0 & 0 & 0 & 0 & 0 \\ 0 & 0 & 0 & 0 & 1 & 0 & 0 & 0 \end{bmatrix}, D = 0_{5\times3}$$

The implementation of the flight control system is shown in Fig. 7.3. Note that both systems implement the same stability augmentation but simulates two different autopilot designs: PID and the Robust Control referred as H-infinity. The original example includes roll-off filters with cut-off at 40 rad/s to partially limit the control bandwidth and safeguard against neglected high-frequency rotor dynamics. To facilitate the control design discussions, these low-pass filters are neglected in the control design discussed below. The low-pass filters are required mostly for practical implementations and their presence is expected to not make much difference at the control design stage. Finally, we have modified the simulations by adding output disturbance signals at the output to assess the disturbance rejection capabilities of the system.

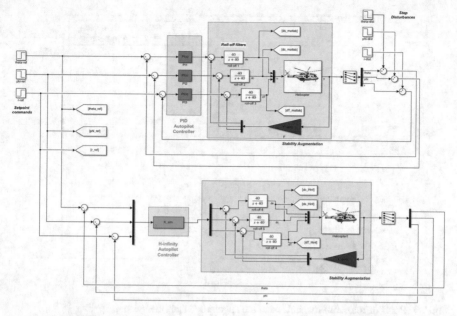

Fig. 7.3 Simulink simulation diagram

7.5.1 Stability Augmentation

Examining the eigenvalues of A, we can observe that the pair $0.0544 \pm j0.4415$ is not stable. In addition the pair $-0.0746 \pm j0.4077$ is poorly damped hence leading to the requirement of implementing first stability augmentation to improve over these two undesirable characteristics. This Matlab example is designed to locate the closed-loop dynamic modes at the following locations

$$\begin{bmatrix} -24.4645 \\ -16.3686 + j3.8325 \\ -16.3686 - j3.8325 \\ -14.4513 \\ -2.7069 \\ -0.3002 \\ -0.0113 \\ -0.0131 \end{bmatrix} \tag{7.39}$$

with the purpose to speed up the response (larger real parts), stabilise the unstable pair (closed-loop poles with negative real parts) and reduce damping (poles with no or small imaginary component relative to its real part), see Fig. 7.4. This example uses the following gain matrix when the stability augmentation system is implemented as Output Feedback

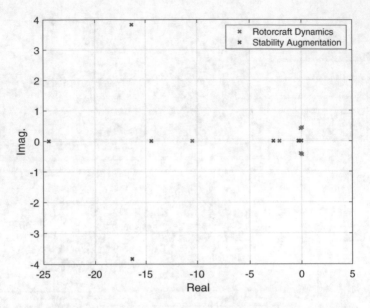

Fig. 7.4 Poles of rotorcraft dynamics and the augmented stability system.

$$F = \begin{bmatrix} [r]10.8486 & -1.7702 & 0.0093 & 1.1523 & -0.1259 \\ -1.1653 & -0.1922 & 0.0193 & -0.1281 & -0.0715 \\ -5.7960 & 2.1839 & -1.0498 & -0.5245 & 0.3714 \end{bmatrix} \tag{7.40}$$

Alternatively, if the scheme is implemented as State Feedback, the corresponding matrix H becomes

$$H = FC = \begin{bmatrix} [r]0 & 0 & 0.5906 & 1.5030 & 0 & 0.0279 & -0.0992 & 0.0350 \\ 0 & 0 & -0.0750 & -0.3683 & 0 & -0.1330 & -1.5900 & 0.0094 \\ 0 & 0 & 0.0325 & 0.0138 & 0 & 0.0588 & -0.0169 & -1.9820 \end{bmatrix} \tag{7.41}$$

The frequency response of the original dynamics and the augmented system are shown in Fig. 7.5. The singular values at each frequency represent the gain variation of the stability augmentation and an additional benefit we observe is that we obtain a lower gain variation especially in the bandwidth region (frequencies less than 10 rad/s). This would facilitate the design of the autopilot system, especially in terms of decoupling the steady-state behaviour.

7.5.2 Autopilot System

In this section we compare two design approaches discussed in Sect. 7.5.2: the conventional PID control which is provided in the Matlab demo and the Robust

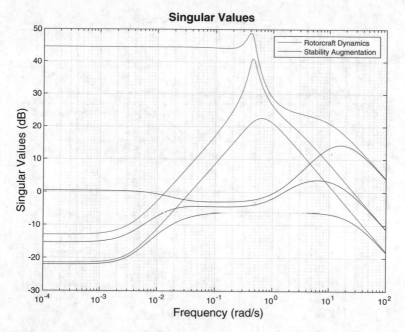

Fig. 7.5 Frequency response of rotorcraft dynamics and the augmented stability system

control design. The PID Control provided in Matlab does not have derivative part ($k_D = N = 0$) and the transfer functions are

$$\text{Pitch:} \quad \frac{2.08}{s} + 1.050$$
$$\text{Roll:} \quad -\frac{1.35}{s} - 0.105$$
$$\text{Yaw rate:} \quad -\frac{2.21}{s} + 0.131$$

For more information on the tuning of these parameters, refer to the Matlab demo.

Note from the Simulation diagram in Fig. 7.3 that the implementation of the PID controller is SISO, hence it is not able to compensate for any couplings in the system. For this main reason we explore the design using the Mixed-Sensitivity design approach to explore if we can get a control design which offer better performance in terms of faster responses, comparable robustness and decoupling.

To achieve decoupling to step commands, we introduce first a pre-compensator based on the dc-gain of the stability augmentation system. Denote the transfer function of the stability augmentation system as

$$\hat{G}(s) = \hat{C}(sI - (A - BF\hat{C}))^{-1}B \tag{7.42}$$

where $\hat{C}$ is obtained by extracting the first three rows rows of the matrix C so the output signal in this case becomes $\hat{y}(t) = [\theta(t), \phi(t), r(t)]^T$. Introducing a pre-

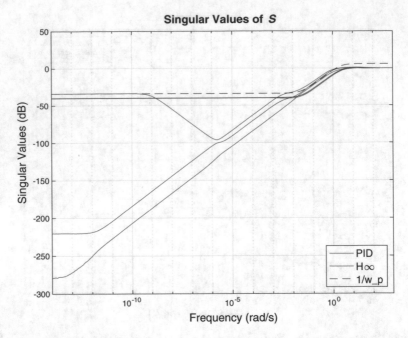

Fig. 7.6 Autopilot sensitivity

compensator $\hat{G}(j0)^{-1}$ would allow decoupling provided $\hat{G}(j0)^{-1}$ exists. In this example this inverse exists because

$$\hat{G}(j0) = \begin{bmatrix} 0.1547 & -0.0633 & -0.0017 \\ -0.0767 & -0.1566 & -0.0030 \\ -0.1318 & -0.8922 & -0.5065 \end{bmatrix} \tag{7.43}$$

is nonsingular. Therefore we can perform mixed-sensitivity design but for the system $\hat{G}(j0)^{-1}\hat{G}(s)$ and thus achieve decoupling of step command signals. We perform mixed-sensitivity control design by shaping the sensitivity transfer function and introducing mild restrictions on the control actions. After some iterations, we choose the performance weight parameters for $w_p(s)$ as follows:

$$M = 2 \tag{7.44}$$

$$S_0^* = 2 \times 10^{-2} \tag{7.45}$$

$$\omega_S^* = 1.5 \, \text{rad/s} \tag{7.46}$$

We choose the control input weight $w_{KS} = 0.15$ because the maximum step command amplitude $\bar{c}$ is assumed to be around 5 and we assume enough control input authority such that $\bar{u} \leq 33 \approx 0.15^{-1} \times 5$ for every control input at steady-state.

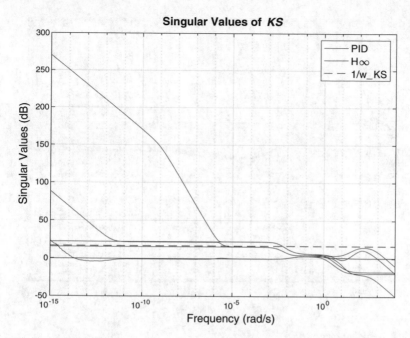

Fig. 7.7 Autopilot control actions

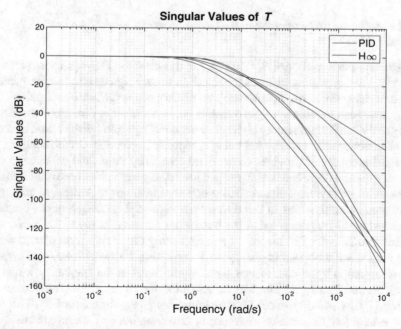

Fig. 7.8 Autopilot co-sensitivity

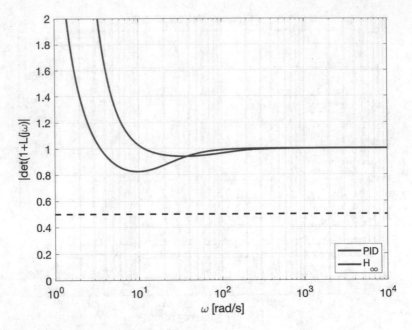

Fig. 7.9 Autopilot robustness

The results are first examined in the frequency domain as shown in Figs. 7.6, 7.7, 7.8 and 7.9:

- Inspecting the sensitivities $S(s)$ in Fig. 7.6, it is clear the PID controller performs poorly when providing disturbance rejection to step signals and low frequencies disturbances. For some combination of disturbance amplitudes and phases, the sensitivity can provide very good results but in some other directions the largest singular value is significantly high hence not securing a good disturbance rejection in all directions and actually not being acceptable in practice. On the contrary, the H-infinity controller provides an excellent sensitivity shape and we can see very small variations between the largest and lowest singular values across all frequencies, ensuring excellent disturbance rejection capabilities regardless of amplitude and phase combinations in the disturbance signals. The design upper bound is satisfied at all frequencies and the bandwidth achieved is $\omega_S \approx 2.4$ rad/s, suggesting sufficiently fast responses and exceeding the design requirement set by ω_S^*. The sensitivity peak is just above 1 suggesting that worst-case performance does not amplify the disturbance signal significantly at the frequency where the peak occurs. Finally, the singular values of the sensitivity at low frequencies are around 0.01 (-40 dB), exceeding the initial design requirement of S_0^*. This very low gain at low frequencies translates in achieving integral action effectively, in other words, total attenuation to step disturbances.
- The H-infinity controller offer significantly lower control input energy when comparing the gain of KS in the low-frequency region between the two approaches,

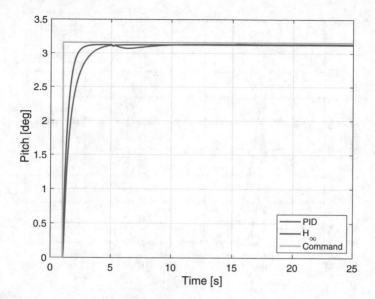

Fig. 7.10 Pitch tracking and regulation

see Fig. 7.7. Large gains to particular directions can lead to extremely large control actions with the PID controller which could not meet actuator or real operation requirements. The H-infinity does not actually meet the initial upper bound requirement for a range of low frequencies so further time-domain simulations would need to be implemented to assess whether these control actions are effectively acceptable or not, and if not, further tuning would be required.
- The co-sensitivity frequency response shown in Fig. 7.8 achieved by both controllers are very good and quite similar. Both designs are excellent in the sense that there is not much difference between the lowest and largest singular values in the bandwidth region of operation, with the gains being very flat and practically 1. The H-infinity controller appears to offer a better co-sensitivity bandwidth around 2.5 rad/s, instead of 0.79 rad/s offered by the PID control scheme.
- The multivariable Nyquist stability criterion implemented in Fig. 7.9 shows that both designs are fairly robust in the sense that they are sufficiently far from the critical point

$$|\det(I + \hat{G}(j\omega)K(j\omega))| \geq 0.5, \forall\omega \qquad (7.47)$$

Around 9.14 rad/s, the above value for the PID controller is closer 0 than the H-infinity hence suggesting that the H-infinity offer slightly better robustness characteristics. This design would suggest that both feedback control methods offer a certain level of tolerance, preserving stable operation in the presence of changes in the rotorcraft dynamics.

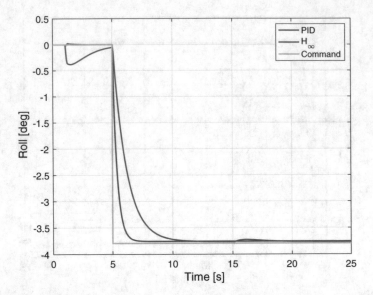

Fig. 7.11 Roll tracking and regulation

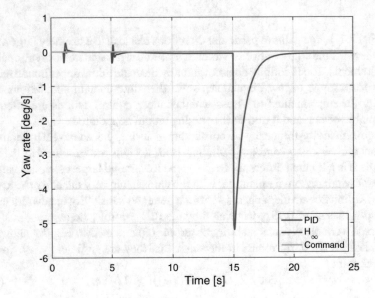

Fig. 7.12 Yaw rate tracking and regulation

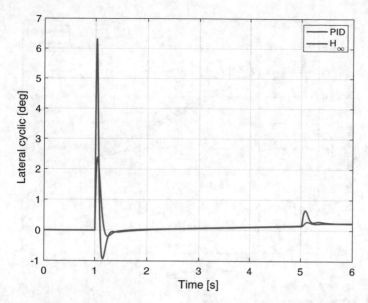

Fig. 7.13 Lateral cyclic control actions

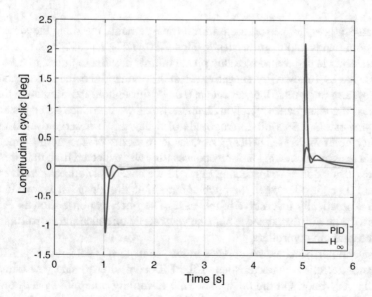

Fig. 7.14 Longitudinal cyclic control actions

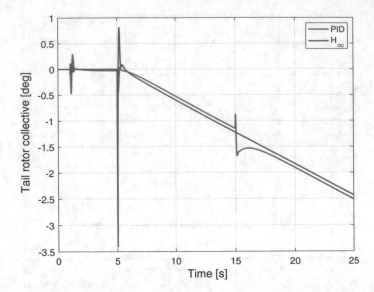

Fig. 7.15 Tail rotor collective control actions

We assess now the performance in the time-domain. We show the simulations to step commands and disturbances in Figs. 7.10, 7.11, 7.12, 7.13, 7.14 and 7.15. The simulation is performed so a step pitch command with amplitude of 3.16 deg is introduced at 1 s, followed by a negative roll step command at 5 s and asking the flight control system to regulate the yaw rate at 0 deg/s throughout the simulation time. To also assess the disturbance rejection characteristics, we introduce a step disturbance on the yaw rate at 15 s with an amplitude of 5 deg/s. We observe overall that the H-infinity controller offer better performance in terms of decoupling among the rotorcraft axes and achieving faster responses (this was expected from the assessment in the frequency domain). Inspecting Fig. 7.11 we observe a noticeable roll response with the PID controller when the pitch command is introduced at 1 s, while the H-infinity is practically insensitive in this case. The pitch response shown i Fig. 7.10 appears to be largely decoupled to both the yaw rate disturbance and the roll command references in both controllers.

The worst performance we observe in the simulation is shown in Fig. 7.12. The disturbance rejection characteristics of the PID controller to step yaw rate disturbances is very poor. On the other hand, the H-infinity autopilot system offers an excellent level of disturbance rejection by keeping the yaw rate close to 0 throughout the simulation time, clearly outperforming the PID controller. As shown in Figs. 7.13, 7.14 and 7.15, cyclic and tail collective control signals performed by the H-infinity controller are much larger in magnitude than the PID controller during the transient, fitting the original control input constraints.

7.6   Concluding Remarks

We have discussed in this chapter state-of-the-art control design methods and have applied them to the flight control design problem of a conventional helicopter, demonstrating key benefits in terms of improved performance, robustness and simpler tuning procedures. We did not discuss implementation implications between classical control and robust control methods. Typically, robust control laws incur in higher implementation costs associated with larger memory requirements and additional computational burden. However, such higher costs are expected to pose no limitations in modern engineering applications given the high processing power of existing embedded systems. The presentation of the topics in this chapter around the robust control methods were constructed using a very informal mathematical terminology to facilitate the introduction of the concepts and focus on the benefits of this control strategy. For a more comprehensive and detailed treatment on Robust Control the Reader is referred to the provided references. Finally, we have shown that the control design *always* demands an exhaustive assessment both in the frequency domain and the time domain to have a reliable assessment. For instance, the PID autopilot provided in the Matlab demo could hint at acceptable performance if assessing only tracking characteristics and no disturbance rejections. This is misleading and the comprehensive assessment in the frequency domain highlighted the weakness in decoupling the system and poor disturbance rejection characteristics. Comprehensive assessment campaign of the flight control system is necessary to build very good confidence on the performance and robustness characteristics of the control design and also for certification purposes. In some applications, the performance and robustness benefits offered by advanced control design methods might not justify the additional implementation requirements so it is the task of the flight control design engineer to overweight these conflicting requirements and choose a strategy which achieve a desired trade off among the many conflicting requirements.

References

1. Skogestad S, Postlethwaite I (2005) Multivariable feedback control: analysis and design, 2nd edn. Wiley
2. Brunton SL, Krutz N (2019) Data-driven science and engineering. Cambridge University Press
3. Johnson W (2013) Rotorcraft aeromechanics. Cambridge University Press
4. Franklin GF, Powell JD, Emami-Naeini A (2005) Feedback control of dynamic systems, 2nd edn. Pearson
5. Cook MV (2013) Flight dynamics principles. Elsevier
6. Scherer C Lecture notes on the theory of robust control. Available at https://www.imng.unistuttgart.de/mst/files/RC.pdf
7. Young N (1988) An introduction to hilbert space. Cambridge University Press
8. Horn RA, Johnson CR (2012) Matrix analysis, 2nd edn. Cambridge University Press
9. Doyle JC (1978) Guaranteed margins for LQG regulators. IEEE Trans Autom Cont 23(4):756–757

Rafael Morales is at the School of Engineering, University of Leicester, control engineering and has contributed to rotorcraft control systems.

Chapter 8
Rotorcraft Preliminary Design

Antonio Filippone and George Barakos

Abstract This chapter introduces conceptual and preliminary design concepts for a conventional helicopter. The subject of rotorcraft design is too vast to the presented in a single book chapter, and thus we limit the presentation to a few key design aspects, as discussed in our lecture series. We provide a top level discussion of the multi-disciplinary team-work required, including analysis of design requirements, costs, risks and compliance. The design consists of a main rotor system and the associated tail rotor, both of which require sizing, with determination of the number of blades, rotor solidity and angular speeds. We discuss concepts for rotor blade design, which is essential in reducing shaft power, improving overall rotor performance, limit vibrations and noise. Concepts are shown for fuselage sizing, including assessment of aerodynamic drag We discuss items such as the empennage (horizontal and vertical stabilisers) and the landing gear.

Nomenclature

MCP	Maximum Continuous Power
MTOP	Maximum Take-off Power
OAT	Outside Air Temperature
SEP	Specific Excess Power
TAS	True Air Speed (called V in equations)
SFC	Specific Fuel Consumption
A	area
A_c	blade section area
c	blade chord
C_D	drag coefficient
d_w	wheel diameter
D	aerodynamic drag

A. Filippone (✉)
The University of Manchester, Manchester, UK
e-mail: a.filippone@manchester.ac.uk

G. Barakos
Glasgow University, Glasgow, UK

© The Author(s), under exclusive license to Springer Nature Switzerland AG 2023
A. Filippone and G. Barakos (eds.), *Lecture Notes in Rotorcraft Engineering*,
Springer Aerospace Technology, https://doi.org/10.1007/978-3-031-12437-2_8

k	main rotor induced power correction
k_{fin}	tail rotor/fin interference induced power correction
M	true Mach number
n	number of operating engines
N	number of rotor blades
P	power
q	dynamic pressure
R	rotor radius
Re	Reynolds number
V	airspeed
V_b	blade volume
W	gross or take-off weight, kg
w	wheel width
W_{f6}	fuel flow per engine, kg/s
x_{tr}	longitudinal distance between rotor shafts
z	flight altitude (m or feet)

Greek Symbols

δ	relative air pressure
λ	induced velocity ratio
μ	rotor advance ratio
σ	rotor solidity
θ	relative air temperature
Π	overall pressure ratio
Ω	rotor angular speed
$\overline{(.)}$	mean value
$(.)_h$	hover
$(.)_i$	index counter
$(.)_{tr}$	tail rotor
$(.)_o$	reference condition

8.1 Design Requirements

Aircraft design is about processes and tools. Tools, however accurate, are of little use unless the correct processes are followed to satisfy the contractual requirements, minimise costs and risks. Thus, the final design is the result of several stages, with further iterations at each stage and compromises at all steps. A *system engineering approach* is recommended. This includes affordability, safety and reliability, product support, life-cycle management, training and human resources. Ultimately, the goal is to proceed from an operational need to a delivered capability. This is done by decomposing the top level requirements into a design strategy that then builds up

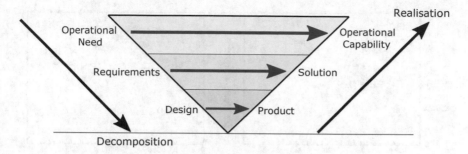

Fig. 8.1 Systems engineering approach to deliver operational needs

into a validated engineering solution and hence to a viable product that delivers the contractual needs, Fig. 8.1.

The history of helicopter design is full of examples where, with the helicopter entirely feasible up to the level of flight testing, the programme was cancelled because of escalating costs, affordability, lack of customers or just obsolete by the time of going to market.

The critical issue of *design choice* is not well captured in a design process. For example, if a numerical value is sought, either continuous or discrete, a numerical model can always be proposed. However, choosing between two different technologies for the same problem cannot always be formalised in mathematical terms. For example: how do you make a high-level choice for a helicopter rotor head or a tail rotor configuration?

Within a sophisticated simulation approach (and there exist many within the research community), we can analyse almost any level of complexity and provide an optimal numerical solution. However, this is unlikely to capture all the constraints, the risks and the certification requirements. *Until the flight test campaign is completed, there are considerable risks and uncertainty.*

The requirements for helicopter design can run into thousands of user needs, top level requirements, and derived requirements. They can further include flight safety constraints, performance guarantee, regulations compliance, certification requirements, and a host of other issues that have to be constantly updated with demonstration of systems and sub-systems compliance.

Conceptual design is the set of engineering operations that allow the definition of the size and shape of the vehicle to fulfill user needs and top level requirements. This is done with simple but robust mathematical methods. Whilst historical data are useful (and are widely used), they inevitably restrict the design within the known parameter space.

The next step is *preliminary design*. In this step we seek to optimise critical elements of the vehicle, and higher resolution simulation methods may be needed at this stage, to produce near-final blade designs, loads estimations, sub-component locations, etc. At this point a guarantee of overall performance and operating costs should be available.

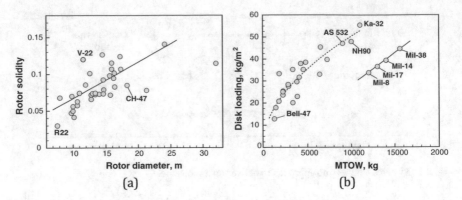

Fig. 8.2 Some statistical data for helicopter rotor solidity and nominal disk loadings

Finally, there is the *detailed design*, when all the geometrical elements are elaborated to their finest detail, materials are chosen, production methods are decided. For this reason, in recent years there has been a widespread use of multi-disciplinary design and optimisation methods [1–3].

Key decisions taken during the conceptual design have the largest effect, the greatest risk and life-cycle cost. These decisions are taken without committing large financial resources, but they are nevertheless critical on the whole design process.

The use of statistical data has some major pitfalls. Whilst it can point to a reasonable initial sizing [2, 4], it can miss completely the optimum configuration; with the modern simulation techniques [5], this is discouraged. The use of statistical data is likely to provide only marginal improvements over an existing rotorcraft, and it will not demonstrate sufficient advantages in a competitive market. There is always the risk of *confirmation bias*, that any experienced scientist and engineer must avoid [6].

An example of such analysis is shown in Fig. 8.2. The rotor solidity appears to be growing with the weight and size of the aircraft, but it is not straightforward to draw a least-square fit, because there are important deviations. For example, compare a heavy tilt-rotor such as the Bell-Boeing V22 and the large Mil Mi-26. These two rotorcraft share a similar rotor solidity, but vastly different diameters and disk loading. Likewise, the design point of very large Russian Mil helicopters would be a total outlier when compared with other general utility helicopters, graph 8.2b. Therefore, an initial starting guess for a novel rotorcraft might not be a good idea after all.

A wide variety of industry data such as those shown Fig. 8.2 is available in the open domain, but also through commercial entities that sell intelligence to concerned organisations. These data are necessary to understand the direction of technology, keep abreast of the competition, and identify gaps in the market.

It is essential to have a historical look at the rotorcraft development, because *unless we are able to demonstrate a measurable step change in critical areas of the market, the helicopter will not sell*. This is a critical junction in the design evolution: since the risks and the costs associated to new rotorcraft are enormous, the industry always grasps between an evolutionary approach of proven technologies, and brand

Table 8.1 Top-level helicopter design specifications

Item	Value	Comment
Helicopter type	–	Conventional
Max take-off weight	3000 kg	
Max payload	800 kg	Internally loaded
Max payload range	250 Nm	
Pax capacity	6	Fully seated
Operating temperatures	−30 − 45	Degrees
Normal Operating altitude	0–3000 m	
Certification Type	–	Category A
Max longitudinal dimension	15 m	
Max rotor diameter	12 m	Limited by ground handling
Max operating speed	–	Unspecified
Noise level		To meet latest regulations by a margin

new concepts that are delivered with greater uncertainly. The conservative approach of the aerospace industry may have settled in spite of enormous progress in design and simulation capabilities. This is one reason why electrically powered VTOL will face resistance on their to new aerospace markets [7, 8].

As discussed in Chap. 4, a key element in rotorcraft performance is the evaluation of direct operating costs (DOC). These are reflected into pounds/dollars per flight hour. Once locked into a design choice, the bulk of these costs would have been forecast. There is considerable amount of evidence in the technical literature that critical design reviews early in the development process lock in operating costs during the lifetime of the vehicle.

With these caveats in mind, we proceed to the preliminary sizing of the rotor systems and the fuselage. Then we demonstrate finer details such as blade design. We provide the key user requirements in Table 8.1. In these top level specifications, the maximum speed is not given, and is left to the designer; hover performance has to be optimised, presumably because the customer may want to do short-distance operations where hover and low-speed is an important phase of flight. Note that a Category A rotorcraft would require a twin-engine configuration. All other parameters are *derived requirements*.

Preliminary sizing of the helicopter is not necessarily a complicated task, and a number of computer codes have been developed over the years to do just that, from the HESCOMP developed by Boeing-Vertol in the 1970s, to modern-day NDARC [1] and several others developed by organisations around the world. In the early days of rotorcraft engineering, design was accomplished with considerable amount of trial and errors. Some of the modern programs are proprietary and others have been published in the technical literature. Features of such programs include the ability to provide parametric studies and/or system and sub-system optimisation. They are generally made of up a combination of models for configuration aerodynamics, weight estimation, structures, propulsion and other elements. These computer optimisation models contain several nested loops that allow iterations from the initial constraints to an aircraft model that is near optimal—at least at this point of the design.

Although initial sizing could be done without major complications, delving into the details of each system, such as the tail rotor and the empennage, is an operation of daunting complexity that may lead quirky ideas, unfamiliar solutions and counter-intuitive compromises. The helicopter never fails to amaze an expert engineer.

8.2 Design of the Helicopter—A Multi-disciplinary Task

In contrast to fixed wing systems, the design of a helicopter is a more integrated task that requires close collaboration between many different departments within a company.

The helicopter is never delivered in isolation. The designers must address pilot skills, recurring training, flight simulation, ground infrastructure, maintenance and overall costs. The knowledge base required to manage these intricate interfaces goes beyond a single discipline, and although multi-disciplinary optimisation methods [3, 9] come in handy, there is no replacement for design experience.

Typically, a chief engineer is responsible for the overall project, and most of the times they will manage a team of specialists. There are many different structures of how the design is organised, but in general the work should include at least four teams to cover the following: rotor designs and rotor-related systems, mechanical design and related systems, structure and materials and avionics.

Although the different teams can work independently, at the start of their tasks there are items that require collaboration even at early stages like for example vibration that is related to the rotor-airframe response. Therefore, the role of the chief engineer is important in managing these interactions and keeping a systems design approach that will deliver the necessary compliance between the complex systems and sub-systems that constitute the rotorcraft.

8.2.1 Rotor Systems Team

The rotor team is responsible for the design, development and testing of the rotors, the hub and rotor shaft. Rotor blades for main and tail rotor, prop-rotors or even propellers for compound helicopters are the responsibility of the team. This is where the inter-dependencies: the rotor is connected mechanically through the drive-train mechanisms to the rest of the vehicle. Thus, vibration and resonances are linking the work of this team with the structures team. There are performance requirements from rotor blades and hubs, and good control margins. Handling qualities are dependent on the design characteristics of the rotor system.

The design of the rotor is not possible without established and validated tools for the two- and three-dimensional computational analysis. The design office should have access to a variety of tools, from low-order first-principle methods that are used to make initial assessments to high-fidelity multi-physics computer codes that require considerable computational efforts and just as considerable professional expertise within the team. The analysis starts from two-dimensional steady state aerodynamic flows and proceeds toward higher levels of complexity that involve wakes interactions, acoustics and aero-elasticity.

The control systems are part of the rotor design and are linked with hub design. Swashplates, actuator jacks, pitch links and horns are part of this effort. The overall geometry and configuration are to be designed, and detailed loading computations are needed for sizing and stressing each member of this assembly.

Tail rotors are an integral part of the rotor systems, although this system has its own peculiarities, as discussed in Sect. 8.5.1. For conventional helicopters, they provide anti-torque to the main rotor, but are also important for yaw control, stability and directionality of flight. Until recently, tail rotor design was not as detailed as main rotors. Nevertheless, the need to reduce the tail rotor power requirements, their importance for noise, vibration and maneuverability resulted in more detailed design in recent years.

An important difference between main and tail rotors is the lack of cyclic control with flap and pitch usually coupled via the delta 3 control angle that links flapping and pitching. Thrust output is the main design objective, but we must also reduce noise and power. Avoiding stall is key, but equally important is the performance with side-slip at different wind conditions are required for certification purposes.

8.2.2 Mechanical Team

The mechanical systems team must deal with engine installation, fuel and transmission lines, hydraulics, undercarriage, and even flotation devices for naval helicopters. Some of the systems used in helicopters are like fixed wing aircraft. Nevertheless, there are significant differences in the hydraulic system used for flight control, and engine installation.

Key parameters that shape the work of the mechanical systems is the location and number of engines, the engine type, power, and acceptable level of vibration. The upper deck above the helicopter fuselage is designed with the mechanical systems in mind. Fuel tanks, their weight, size and location, fuel pumps and fuel lines are among the sub-systems must be addressed by this team.

Helicopter fuel systems design follows similar procedures to fixed wing aircraft tanks. In any case, the tanks are designed to be placed near the aircraft centre of gravity (CG). The emptying of the tanks is sequenced to smooth changes in the CG. Modelling fuel systems on a computer is now possible, although full-scale tests on ground rigs are necessary. Data from such tests, as well as the flight trials, are part of the certification process. Even full-scale mock-ups of the overall system are used to ensure that all sub-systems fit together.

Looking even at one specific issue such as the engine Chap. 3, it is enough to see the impact of the work of the mechanical systems in the overall design. The engine will have to be placed on top of the airframe; due to weight of the engine and the gearbox, there is a need carefully manage the placement of several components. Alongside the engines and their start-up system, the gearbox needs a support structure with vibration absorption. Special care needs to be taken for the main and tail-rotor output shafts and drives. Sometimes ancillary drives are required and there is a need for a rotor brake, firewalls, intakes and even monitoring of the level of pollutants on the *upper deck* of the helicopter. There are several systems with high temperatures requiring fire suppression systems, management of materials to allow for temperature gradients to be smoothed out and controlled. The intakes and exhausts of the engines are important design tasks. Experience in design will allow the team to consider accessibility, delivered through access panels and removable cowlings, maintainability, and easy inspection.

The transmission is the heart of a helicopter system. As with any critical system, modes of failure and ways to mitigate these become part of the design. Gearbox operation at high temperature without oil is always a risk to consider; the aim is not to develop a binary system that works or fails, but a system that degrades its performance gradually before failing completely.

Detailed finite-element computations are common to determine the safe sizing of the teeth, gears, and casing of the gearbox. This analysis is often coupled with thermal stress calculations and even CFD for the flow inside the upper deck is common to ensure cooling. The design of such a critical system is linked with the Health and Usage Monitoring System of the aircraft (HUMS) this is important, since HUMS can gather data and provide guidance for the required maintenance of each system. There is a wealth of information gathered by HUMS and data mining algorithms are used to deliver info related to scheduled or ad-hoc maintenance of each component. It shows the strong link of the mechanical systems with both rotor systems and avionics teams.

Some of the most stringent tests are required as part of designing and proving of this system. The rotor brake, driving shafts, pressure lines will have to be tested to failure. The engine and its control are part of the test. With rotor shaft rpm in the range of 200–280 and engines outputs near 25,000 rpm, there is a need for staged reduction on the engine and gearbox or on stages inside the main gearbox Chap. 3. Engine manufacturer and mechanical team are to work together to finalise the design.

APU/starter integration with the engine is required on many rotorcraft. Some helicopters require a ground-based power supply, but this is seen as a limitation, and batteries can be used if the installation of an APU is not feasible. The engine and its Full Authority Digital Engine Control (FADEC) must be integrated with the aircraft system.

The wheeled undercarriage design shares similarities to fixed wing aircraft applications, although considerations of phenomena like ground resonance are still very important. Ground resonance has resulted in several helicopter losses and there are spectacular images of the level of distraction it brings. Careful dynamic design and good theoretical understanding are used to alleviate this problem. Most undercarriage design is carried out in collaboration with an external contractor.

Crashworthiness and post-crash survival rates are essential in the certification process. The objective is to maximise the amount of load taken in a heavy landing without propagating damage. Good collaboration and communication are key to the success of the design and to minimise costly mistakes. Some helicopters must consider flotation systems, which are manufactured by independent contractors.

Helicopters have a difficult CG distribution with the engines mounted high on the airframe; therefore, there is a need for stability and safety when flotation bags are deployed. Even if there has been significant progress with simulation of such systems, there are practical tests at model scale and full scale required to assess *helicopter ditching* early in the design.

8.2.3 Structures and Materials Team

The structures and materials design team should deliver the design of the main airframe structure with the associated loads estimates as well as provide data for sub-structures, for example doors, hatches, panels, transparencies. The team works with the mechanical systems team closely since some loading aspects, for example ditching loads, undercarriage loads, thermal loads must be estimated. The metal structures that dominate helicopter airframes are recently challenged using composite materials. For the NH-90 helicopter design a significant use of composites on the airframe was made. This approach was backed up by substantial research to deliver on the requirements of such a novel structure.

The loads on the structure are divided in static and transient. Static loads are further sub-divided in flight loads and landing loads. Transient loads are associated with the longevity of the structure and fatigue. Apart from this crude sub-division, there is a need to ensure that the structure is free from mechanical resonance and that the natural frequencies and modes of vibration are not multiples of the rotor forcing frequency. Furthermore, the crew and cabin space are designed for maximum survival rate in a crash. Military helicopters have further requirements for ballistic strike.

Most of the analysis is now carried out using simulation tools. These are dependent on accurate material properties and on models to allow for simulation of contacts, rivets, joints that are difficult theoretical problems under investigation. Crack and damage propagation models are used to perform detailed structural analyses. There is a need for additional modelling and simulation to allow for composite materials to be analysed, and consider effects such as weathering, resistance to salt, corrosion, and electromagnetic compatibility. Yet, the final certification will need large parts of the structure to be statically tested on the ground.

Dynamic tests are required; these tests must demonstrate the ability of the structure to sustain periodic loads for the planned lifespan of the structure. Good communication and iterative design are needed with the rotor systems team to allow for avoiding resonances, minimising vibration, and delivering a solid design that can meet the current requirements but also has some growth potential.

8.2.4 Avionics and Electrical Systems Team

The avionics team is getting more and more prominent role in modern helicopter design. The helicopter is a platform of sensors, systems, and tools alongside a transport for humans and goods. For this reason, the overall design is getting more linked with the avionics systems on board. We already touched on the importance of HUMS and how these may be integrated with the design process.

Electric power is needed, and this usually covers AC and DC supplies to different systems. The electrical system should aim for a redundant power supply, and batteries are also used. Given that most helicopters have accurate governors that maintain the shaft speed constant to 1–2% of the rated rpm, the power supply should be fairly stable. The use of simulation tools to size and design the electric system, as well as simulate its function are becoming more and more common. Furthermore, there is a strong need for electro-magnetic shielding to avoid electronic interference. This may be especially true for systems on the helicopter that consume significant power, for example de-icing, operation of heavy winches and electric doors and ramps.

The electric supply will also feed the avionics suite of systems that are required to operate the aircraft. Most helicopters are nowadays complex electronic systems where data and information management are taking place continuously. Therefore, there is a tendency for helicopters to be fitted with data-buses that share and distribute data with the support of central processing units. Multiple buses are accommodated, particularly for military helicopters with a variety of weapon systems.

A relatively recent trend is the use of the so-called open architectures with modular avionics units that are integrated with the airframe via sensors, data, and electric buses. Alongside HUMS, there are aircraft that have additional test equipment that gather data from different flights and operations to support the design of future helicopters and to improve operations and maintenance.

The cockpit of the helicopter is designed balancing functionality, ergonomics, weight, and maintainability. Even if it is one of the responsibilities of the avionics team, there is strong need for multi-disciplinarity, and all teams must provide input. As part of the design effort, anthropometric data is used centering the design around the human pilot who will have to have enough reach to operate the various systems. Emphasis is placed on the crew physiology; temperature control and ventilation are needed to support the pilots work in what is a rather confined space with several demands.

Given the importance of night operations or operations at degraded visual environments (DVE), there is a need for non-visual gauges, audio cueing, as well as, illuminated high quality displays. All this effort aims to reduce the workload of the crew. At times, a mixture of electronic and analogue/mechanical gauges are needed to allow for redundancies and safe operation of critical systems. With two or sometimes three or more people needed to fly the helicopter and operate its systems, there is a need for multi-crew collaboration, separation of responsibilities and tasks and this calls for certain displays and systems to have duplicates and duplicate controls. Managing the information and prioritizing the display of information is a key task in the modern design of the cockpit. Amount of information and order of information provision are strongly related to the workload of the crew that must share different tasks and perform different operations at the same time. Customisation is often provided on request.

Multi-Function Displays (MFD) tend to have a good level of customization and serve as the main source of information. The display area generally is below the windscreen and between pilot and co-pilot. Additional panels can be found centrally between the crew, but there are also head-up displays and on-helmet displays. The inceptors (cyclic and collective) are frequently equipped with more switches and jogs to allow for several functions to be performed from there.

8.3 Analysis of Requirements

The large set of requirements for a helicopter design must be dissected into categories that can be analysed at a top level. We review the requirements on mission specifications, overall dimensions and rotorcraft weight. In Chap. 4 we presented aspects of direct operating costs and payload-range performance. These concepts are very important at the conceptual design stage. They will reflect into a unit cost that the customer will have to pay for each flight hour.

8.3.1 Overall Dimensions

Size limitations are a fact of life for any engineering vehicles, from cars to airliners. Helicopters are no exception. Longitudinal dimensions are interpreted as *rotors running*; for a conventional helicopter, lateral dimensions are equivalent to the maximum diameter; the vertical dimension is limited either by rotor running or by the fin. The fuselage, discussed separately, is enveloped by these dimensions. The overall dimensions dictate where the helicopter can operate, and how close it can get to buildings and obstacles. Some helicopters have folding blades that can be stowed to reduce space requirements on the ground (for example, the Leonardo EH101/AW101, the Airbus EC145/H145, the Sikorsky MH-92). The folding mechanism may become part of the design requirements.

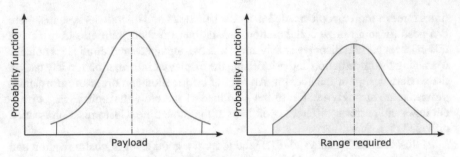

Fig. 8.3 Statistical analysis of mission requirements for conceptual design

8.3.2 Mission Requirements

Mission analysis must be carried out with appropriate statistical assumptions: range, payload and atmospheric condition are rather variable. An example of probability distribution for range and payload is shown in Fig. 8.3. Although the central values are the most likely, we need to consider how the helicopter would operate in extreme conditions.

In a design contract there will be clauses for *performance verification and guarantee*, which will have to be mutually agreed. Performance is normally guaranteed by flight testing; performance prediction by simulation models, no matter how advanced, is for indication only. The contract should also stipulate the details of the flight testing campaign. In Chap. 3 we discussed the case of engine performance guarantee. A good discussion of this aspect of rotorcraft design is available in Stepniewski and Keys [10] (Appendix A). Standard statistical methods are used to establish the most likely performance parameters out of a distribution of measurements.

8.3.3 Rotorcraft Weights

At the concept design level weight estimation remains largely empirical, and based on past experience. Stepniewski and Keys [10] (Volume 2) addresses conceptual design in broad terms and continues to be a good reference, although a number of more academic publications are available as well [5]. The estimated gross weight is

$$W = W_e + W_f + W_p, \tag{8.1}$$

with contributions from the empty operational weight, the fuel and the payload, respectively. The target payload is specified by the customer, Table 8.1. The remaining two terms are unknown. The fuel capacity will have to be determined with typical missions, once a power plant is selected. The operating empty weight is a more serious undertaking. Again, one is tempted to use statistical data in order to establish a starting point. In the process of building up a rotorcraft model, some weight components become available, albeit in approximate value. Hence, by using the principle of components, a structural weight can be estimated. The technical literature reports data that are now quite old and have not been updated for several decades. Therefore, when advanced materials are used, past reference data may be obsolete.

A gross weight is not necessarily specified, hence it is a problem for the engineering team to establish the operational and regulatory weights, most of which have been previously defined in Chap. 4. As in most of engineering design, this is an iterative procedure that involves a first estimate of a structural weight W_e, a required maximum payload W_p and the minimum fuel capacity W_f that allows the rotorcraft to fulfill the specified performance, Eq. 8.1. The most straightforward component is the payload W_p, since this is a top level requirement (Table 8.1). The operational empty weight (OWE) must be established with the method of components, by adding all the sub-systems.

Assume that the OWE has been estimated at some point in the design chain. Hence, we have the zero-fuel weight ZFW $= W_e + W_p$. For a specified mission, calculate the fuel required, including the regulatory reserves. This process will give a first estimate of the fuel capacity required. Example: the rotorcraft has to travel 250 n-miles with the design payload (Table 8.1) in the most adverse weather conditions (hot day, head wind, at a suitable flight altitude). There are several ways of getting the fuel capacity right; the numerical method we use is based on the iterative bisection procedure:

1. The estimated fuel capacity will be within the range $[W_{fmin} : W_{fmax}]$.
2. With the fuel capacity $W_f = 0.5 \left(W_{fmin} + W_{fmax} \right)$, perform a mission analysis
3. If the fuel capacity is insufficient, set $W_{fmin} = W_f$, otherwise set $W_{fmax} = W_f$
4. Restart from point 2 until convergence, which should occur after 3–4 iterations.

Once the fuel capacity has been established, we need to assess the maximum take-off weight (MTOW), which has been given as part of the project specifications. As demonstrated previously in Chap. 4, we cannot operate the rotorcraft with both maximum payload and maximum fuel capacity, there will be trade-offs to consider. The payload ratio is $W_p/$ MTOW ~ 0.27, in line with helicopters in similar weight class. This leaves $\sim 2200\,$kg between OWE and fuel. These two parameters are interdependent; however, as the OWE increases, so does the fuel required to fulfill the contractual missions. Hence, structural weight minimisation is central to the whole design process.

8.4 Main Rotor Sizing

A rotor can be sized and optimised for one or more flight conditions. For example, an optimal rotor for fast level flight will be different from a rotor designed for the best hover/low-speed performance. When more than one operational point is to be considered, then the design must proceed through a post-optimal analysis that evaluates the trade-off between different targets.

The fundamental design parameters of a helicopter rotor are: rotor diameter, disk loading, tip speed (or Mach number), rotor solidity, number of blades, twist distribution, airfoil sections, tip shape, collective pitch range, moment of inertia, and direction of rotation. Thus, we have a design space with about a dozen parameters.

In order to size the main rotor we consider the power equation derived from aerodynamic theory:

$$P(z, V, W) \simeq P_h \left(\frac{\mu}{\lambda_h} \frac{D}{W} + \frac{\lambda_h^2}{\lambda_h^2 + \mu^2} \right) + \frac{1}{8} \sigma \left(1 + 3\mu^2 \right) \overline{C}_D \rho A (\Omega R)^3 \quad (8.2)$$

with the hover power defined by

$$P_h = \frac{k}{R} \frac{W^{3/2}}{\sqrt{2\rho\pi}} \quad (8.3)$$

Note that D in Eq. 8.2 is the *helicopter drag*. The ratio D/W is a quantity that must be estimated with other methods, but it is useful make an initial estimate. Using past data, we find $D/W \sim 1/6$ in level flight, possibly less, for example 1/7. However, this value depends on the flight speed, and would be zero in hover. We need to add a number of constraints. One is on the limit tip Mach number and the other is the maximum practical radius (or diameter)

$$M_{tip} = \frac{\Omega R}{\sqrt{\gamma R T}} \leq 0.85, \qquad R \leq R_{max} \quad (8.4)$$

In Eq. 8.2, $P(z, V, W)$ the rotor power is dependent on the set of operational parameters altitude, speed and gross weight. The free parameters are the angular speed Ω, the radius R, the blade solidity σ and the mean drag coefficient of the blades at the specified flight condition. Since this drag coefficient is to be minimal in all cases, it is more of a matter of airfoil design, and in the first instance it can be removed from the set of thee parameters. The blade solidity is proportional to the number of blades and their mean chord. This is an important design parameter, but it will be split from the initial set of requirements. Therefore, we have

$$P(z, V, W) = f_1(\Omega, R) + f_2(\sigma, \overline{C}_D) \quad (8.5)$$

Although Eq. 8.2 is an algebraic equation, function minimisation using total derivatives is not straightforward, because of its implicit nature. However, parametric studies of optimal radii pose no difficulty. Such a parametric analysis is shown in a 3-D space with the rotor solidity, the mean drag coefficient and the rotor rpm, Fig. 8.4.

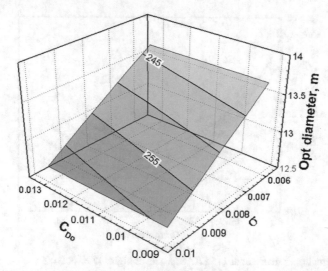

Fig. 8.4 Parametric analysis of optimal rotor diameter

Let us consider a few typical cases, which for the sake of comparison are calculated at the same gross weight, altitude and atmospheric conditions. The reference maximum weight is $W = 3000\,\text{kg}$ (~ 30 kN) at sea level on a standard day.

Rotor Sizing for Optimal Hover Performance. This is the case of a rotorcraft that is specifically designed to operate at low speeds, hover for some length of time, lift and deliver large external (sling) loads with high precision. The rotorcraft may have to operate at high altitudes. In this instance, $V = 0$, $\mu = 0$, hence $D = 0$ in Eq. 8.2. To find a minimum power, we can either do a parametric study or derive the equation with respect to the radius, with other variables in parametric form. In the former case, the radius yielding the minimum power is

$$\frac{\partial P}{\partial R} = -\frac{k}{2}\frac{W^{3/2}}{\sqrt{2\pi\rho}}\frac{1}{R^2} + \frac{5}{8}\pi\sigma\overline{C}_D\rho\Omega^3 R^4 = 0 \tag{8.6}$$

$$R = \left[\frac{4kW^{3/2}}{5\pi\sigma\rho\overline{C}_D\Omega^3\sqrt{2\pi\rho}}\right]^{1/6} \simeq 0.6831\left[\frac{k}{\sigma\rho\overline{C}_D\Omega^3}\right]^{1/6}\left(\frac{W}{\rho}\right)^{1/4} = f(W,\rho,\sigma,\overline{C}_D,\Omega) \tag{8.7}$$

Eq. 8.7 is the *exact* solution of an *approximate* equation, and contains several operational and design parameters. It is noted that for all other parameters being constant:

- As the rotorcraft becomes lighter, the optimal radius increases.
- As the rotorcraft flies higher, the optimal radius increases.

A number of constraints can be introduced. For example, the tip Mach number is $M_{tip} = 0.55 - 0.60$. Larger tip Mach numbers would be encountered in level flight, so this is a limit. Thus, $M_{tip} = \Omega R/a$ and Eq. 8.7 becomes

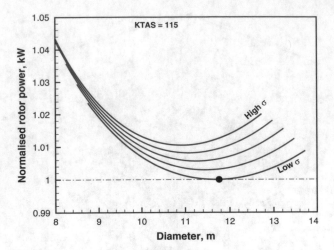

Fig. 8.5 Optimal rotor diameter at 115 KTAS, sea level flight, $W = 3000\,\mathrm{kg}$

$$\frac{c_1}{R^2} + c_2 R = 0, \qquad c_1 = -\frac{k}{2}\frac{W^{3/2}}{\sqrt{2\pi\rho}}, \qquad c_2 = \frac{5}{8}\pi\sigma\overline{C}_D\rho a^3 M_{tip}^3. \qquad (8.8)$$

Rotor Sizing for Fast Level Flight Performance. This is the case of a rotorcraft that is specifically designed to fly fast, operate in hover in very limited circumstances (aside from take-off and landing), and deliver internal loads. The limit tip Mach number on the advancing blade is $M_{tip} = 0.85$. Figure 8.5 shows the parametric effects of the blade solidity on the rotor power, and is useful for preliminary sizing of the rotor in medium to high speed. Note that only an average drag coefficient has been used, with the understanding that drag coefficient minimisation will be undertaken during the airfoil selection process. The power data are normalised to the minimum power of these calculations. It is noted that the optimal diameter at point M is well within the specifications given in Table 8.1; therefore, this is a good starting point.

Blade Sizing. Once the optimum blade diameter has been determined by using a full sweep of design and operational parameters, there remains to define the number of blades and their average chord. The results indicate that low solidity is preferable. However, this is just an aerodynamic result that does not take into account the actual blade loading. This is defined as C_T/σ. The design C_T will be used and a limit blade loading assigned on the basis of past experience and/or further design consideration. Thus, σ is derived from an algebraic consideration.

$$C_T/\sigma = (C_T/\sigma)_{design} \qquad (8.9)$$

with $\sigma = cN/\pi R$. Now we have two parameters (chord c and blade count N), one of which (the blade count) must have discrete values. At the two extremes we have either a rotor with many slender blades, or a rotor with a few chunky blades, because

$cN \simeq$ constant. Normally, there is little freedom at this point, and experience comes in to help, although again one can use sophisticated simulations to get a rational solution.

Structural Considerations. The blades weight is proportional to the number of blades: $W_b \sim N V_b$, with V_b the blade's volume, $V_b \simeq R A_c$, and A_c the cross-sectional area of the blade. Thus, $W_b \simeq N R A_c = N R A_{c1} c^2$, where A_{c1} is the cross-sectional area of a blade with unit chord, defined as

$$A_{c1} \simeq 4.128 \cdot 10^{-3} + 0.645(t/c) \tag{8.10}$$

This cross-sectional area changes little with different blade thickness. Therefore, the rotor weight with the optimised diameter grows with the factor $c^2 N$. For a given average thickness and fixed radius, the total blade weight increases with the number of blades (because $c < 1$, $N > 1$).

A light blade is not necessarily a good thing, because we need to consider the autorotational performance discussed in Chap. 4: a heavier blade has a stabilising effect on the rotor dynamics and stores sufficient energy to allow some flight control in case of loss of engine power. As in previous cases, the solution is no longer straightforward when we consider all the interdependent aspects of the rotor.

8.4.1 Rotor Head Architecture

The rotor shape and dimension are only one aspect of the rotor system design. An important consideration is about the nature of the rotor head, which can be fully articulated, semi-articulated, rigid, etc. Decisions about the rotor architecture cannot be taken on the basis of simulation tools alone, but have to rely on considerable engineering knowledge that cannot always be translated into a computer code. As it often happens, using known technology bears minimum risk, though not the best solution all the time. The technical literature is silent about the design and optimal choice of helicopter rotor heads. In this instance, we assume a rotor head architecture of the same type as other helicopters of similar size. An articulated rotor will have a number of sub-systems to consider, starting from the swashplate.

For an articulated rotor, the selection of offsets between the blade hinges for lead-lag, flap and pitch, and the type of the hub (rigid, semi-rigid, articulated) will have a major impact on the control responsiveness and handling of the vehicle. Such characteristics are important for the control system design. Specific issues related to the hub, such as blade sailing, are important.

Above all, the engineer must be concerned about the number of moving parts, the identification of parts most subject to high loads and fatigue, minimised weight and design fail-safe structures capable of operating under extreme flight conditions. Articulated rotor heads, rotor heads for coaxial helicopters and large helicopters are known to have high drag. Fairings and streamlining are used when possible [11], with some modern helicopters showing very sleek aerodynamic designs (for example, the EC-135).

8.5 Tail Rotor Design

In the previous section we presented a simple method for main rotor sizing. No consideration was done as to the presence of a tail rotor, and the additional power requirements: that contribution would have to be added to the analysis on the next iteration. We consider the case of a conventional open rotor having a constant angular speed, with fully reversible pitch. The design parameters include: rotor location, direction of rotation, diameter, tip speed, rotor solidity, disk loading, airfoil sections, number of blades, twist distribution, cant angle, and collective pitch range. Again, we have at least a dozen different parameters.

Before we begin, we need to consider the most demanding condition, which is the highest hover altitude, at the highest weight, in unfavourable atmospheric conditions. In other flight conditions, the tail rotor is off-loaded with a cambered fin and asymmetric vertical stabiliser. If the hover thrust is guaranteed, then forward flight thrust is also acceptable. Finally, we also need to consider future rotorcraft upgrades. Typically, these are achieved by in increase in certified MTOW, requiring an engine upgrade. An increase in design torque requires an increase in tail rotor torque reaction, hence net thrust, which may not be delivered in all flight conditions. Some guidelines have been established in the engineering literature [12].

8.5.1 Tail Rotor Concepts

Over the years several tail rotor architectures have been invented and applied to real life helicopters. The *conventional* open rotor remains the most widespread example of tail rotor design. Alternatives include ducted tail rotors (for the example, the French Fenestron® [13] on the EC145 and EC365 and prior embodiments SA341, as reported in Ref. [14]); and the jet propulsion systems using lateral airflow (for example the McDonnell-Douglas NOTAR system on the MD-600 [15]).

The conventional rotor can be mounted in-line (at the end of the tail-boom: MD-500E) or *cascading* (on the vertical fin: Boeing AH64, Sikorsky UH60, and S61); it can rotate bottom-up or top-down. Its blades can be equally spaced or can be out of phase. From a propulsive point of view, the tail rotor can be *tractor* (Leonardo AW 139, Sikorsky UH60) or *pusher* (Westland Lynx). It can be placed on the vertical plane (most helicopters) or canted. The CH-53E has a tail rotor disk canted 25°C downward on the vertical plane (though another version, the CH-53D has vertical tail rotor); likewise, the UH-60 has a tail rotor canted upwards 20°C, presumably to decrease its noise signature. Therefore, there are several degrees of freedom.

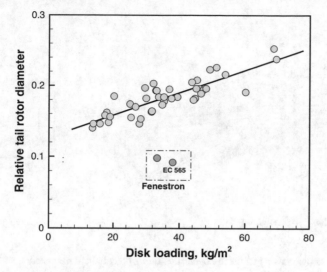

Fig. 8.6 Tail rotor diameter ratio data

Tail rotor diameters are best assessed in relation to their corresponding main rotor. The tail-to-main rotor diameter ratio d_{tr}/d grows with the disk loading, as indicated in Fig. 8.6. If we maintain a similar advance ratio between the main- and the tail rotor, then the tail rotor angular speed is found from

$$\frac{\Omega_{tr}}{\Omega} \simeq \frac{R}{R_{tr}} \tag{8.11}$$

From the data in Fig. 8.6, with a diameter ratio of ~ 0.2, we find that the rpm of a tail rotor is typically five times larger than a main rotor. This is in fact, not far from true. However, we need to clarify that the advance ratio of the tail rotor may need to be sensibly lower than the main rotor, since its blade design tends to be slightly less advanced.

Example of data include the following: for the AS332 Super Puma, the tail rotor has rpm $= 1284$, leading to $\Omega_{tr}/\Omega = 1284/265 \simeq 4.845$. The ratio between rotor diameters is $d_{tr}/d = 3.05/15.08 = 0.206$, which basically satisfies Eq. 8.11. For the Bell 206, it is slightly different: we have $\Omega_{tr}/\Omega = 2550/394 \simeq 6.47$, which is quite large in comparison with other helicopters.

The tail rotor is a peculiar mechanical system, with some problems that need further consideration [12, 16]: these include the gyroscopic effects, the fin interference, blade vortex interaction, and pitch-flap coupling.

Yaw Rates. The tail rotor is a gyroscope that is spinning rapidly and resists changes in the orientation of the shaft (*precession*). When the rotorcraft is in yaw, it increases the out-of-plane velocity component on the tail rotor disk by an amount $\dot{\psi}x_{tr}$, as illustrated in Fig. 8.7.

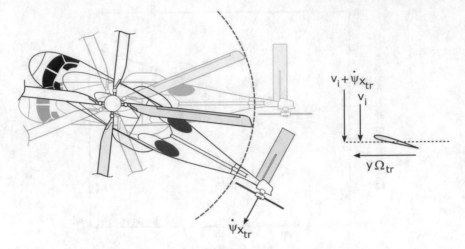

Fig. 8.7 Helicopter yawing whilst in hover, illustrating potential tail rotor stall

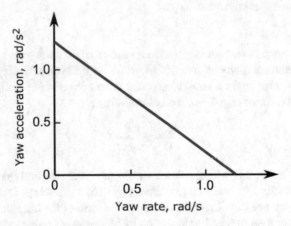

Fig. 8.8 Typical tail rotor stall boundary

The total inflow would be $v_i \pm \dot{\psi} x_{tr}$, with v_i dependent on the tail rotor loading and the direction of yaw. The effective inflow can stall the blade, which leads to a loss of tail rotor effectiveness, with potentially serious consequences. Left-right yaw limits are different; they depend on the yaw acceleration and on the sense of rotation of the tail rotor. A stall boundary is defined for each tail rotor at OGE hover ceiling [16]. A typical stall boundary is shown in Fig. 8.8. The stall boundary of a shrouded rotor (including the *fenestron*) is possibly narrower than the one illustrated below.

Tail Rotor to Fin Interference. Tractor and pusher tail rotors operate differently: the wake strikes the vertical fin in the former case, but not in the latter. This interference is strongest in hover. For a tractor tail rotor in hover, the side-force from the fin causes an increase in net rotor thrust, which at the design tail rotor torque can be as much as 1520%.

For a pusher tail rotor in the same flight condition, interference is accumulated in the rotor inflow. Aerodynamic performance is known to depend on the rotor-fin axial separation; it reduces from 1520% of net rotor thrust at small separations to a negligible loss when this distance is equal to the rotor radius (not a practical solution, in any case).

We need to make consideration on the number of blades of the tail rotor in order to minimise blade vortex interaction and positive interference of tonal noise components. For the AS332 helicopter, $N = 4$, and $N_{tr} = 5$, which correspond to a blade passing frequency BPF of 17.7 Hz 107 Hz, respectively. The frequency ratio $107.0/17.7 = 6.05$, which should avoid the amplification of tonal noise components Chap. 6.

8.5.2 Tail Rotor Sizing

As in the case of the main rotor, we need to establish the power required by the tail rotor to guarantee lateral control in the most demanding flight condition A first-order estimate of this power is

$$P_{tr} = \frac{k_{fin}k}{\sqrt{2\rho A_{tr}}} \left(\frac{P_{mr}}{\Omega x_{tr}}\right)^{3/2} + \frac{\pi}{8}\sigma_{tr}\rho\overline{C}_D(1 + 3\mu_{tr}^2)\Omega_{tr}^5 R_{tr}^3 \qquad (8.12)$$

where the tail rotor solidity is $\sigma_{tr} = (Nc/\pi R)_{tr}$. Equation 8.12 contains a fin interference factor k_{fin} and an induced flow factor k that has the same meaning as in the main rotor. The term x_{tr} is discussed in Chap. 4, where we discuss lateral stability concepts. In this instance, we note that there is limited flexibility on the rotor-to-rotor distance x_{tr}, which is dictated by the overall dimensions of the rotorcraft, weights, structural loads, transmission system and aerodynamic interference, not necessarily in this order.

One specific feature of tail rotor design is the potential interference between the rotor and the vertical fin, which is inevitable. However, this interference depends on whether the tail rotor is a tractor or puller; it also depends on the flight condition. Figure 8.9 shows a drawing of the Eurocopter AS332 *Super Puma* empennage and tail rotor. The rotor is mounted vertically on the opposite side of the fin, and the horizontal stabiliser is mounted high behind the fin. In this case, the tail rotor is a *pusher*. This means that in hover the rotor downwash through the rotor disk would be moving toward the fin.

In forward flight, the tail rotor wake has a skew angle, and the interference diminishes as the rotorcraft flies faster. This is represented by a behaviour expressed by the following equation:

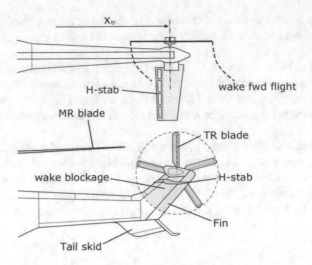

Fig. 8.9 Tail rotor configuration of a Eurocopter AS332. A five-bladed rotor, articulated on the flapping plane, rotates clockwise in the vertical view

$$k_{fin} = \begin{cases} 0 & \text{if } \mu_{tr} \geq 0.05 \\ 1.2 - 24\mu_{tr} & \text{if } \mu_{tr} < 0.05 \end{cases} \tag{8.13}$$

The airflow around the tail rotor blades is intrinsically unsteady, even in hover. This is due to the fact that the close proximity between rotor blade and the hard surface of the fin generates a virtual *ground effect* which translates in periodic blade loadings.

The optimal tail rotor radius from Eq. 8.12 depends on the flight conditions. For this reason, we show a parametric analysis in Fig. 8.10, as function of the rotor solidity for a fixed flight speed and mean blade drag coefficient. The rotor power is normalised with its minimum in order to better show the sensitivity to blade solidity and blade radius. The changes in power are a small percentage, and hence we demonstrate that the initial solution is something we can work on. It is further noted that the rotor rpm was fixed.

8.6 Blade Design Concepts

We discuss the aerodynamic design of the blade, but a separate discussion would be needed for the structural design. With a few exceptions indicated in Chap. 1, rotor blade chords are constant. They may include a control tab and sophisticated tip design [17] to improve aerodynamic performance.

The outside shape is defined by a series of airfoil sections, whose selection is a matter of detailed aerodynamic design aimed at providing the best performance over a full range of Mach numbers and inflow conditions, both steady and unsteady. For this

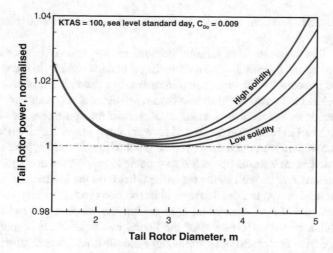

Fig. 8.10 Optimal tail rotor radius at fixed advance speed and mean blade drag coefficient

purpose, there is a vast literature on airfoil design and a variety of high-performance rotorcraft airfoils. These airfoils are supercritical wing sections, for example Sikorsky SC-1094, SC-1095 [18]. In any case, rotor blade sections are selected from a database of aerofoils that are carefully designed to fit the root, middle or tip stations of the rotor. This database is backed up with a large amount of aerofoil performance data for different conditions, checked and cross-checked between different wind tunnels and even CFD computations. The main objectives are to maximise the lifting performance in thrust and power and do this across the expected envelope of operations of the vehicle. Alongside the performance requirements, the design will have to consider the expected noise of the system, means to control vibration, integration with engines and autorotation.

Being the heart of the helicopter, the blades carry the weight and are subject of extremely large unsteady loads. Structural design to minimise vibration and noise is equally important. The blade will have a main spar, a foam filler, a composite construction with several layers and ancillary systems such as sand erosion protection at the leading edge.

Carbon composite blades are the dominant construction, with a central D-spar made out of carbon laminates, foams of different types are used to reduce weight and multi-layered carbon skins cover the blade. Metal or other erosion shields are used. The blades are attached to a hub. Designed to connect the rotor blades and output shaft of the main gearbox, the hub will transmit torque and will provide the necessary loads paths for all forces and moments.

Conflicting requirements between soft blades that untwist in forward flight and stiff blades in bending to benefit blade sailing are not uncommon. Accurate kinematic and dynamic models of the hub are also very important and these can be made using tools developed in-house in various rotorcraft companies or built around commercial tools.

Systems like ice protection, folding mechanisms for blades, lighting strike protection are to be considered. Boundary conditions and restrictions to the work of this team can be imposed by the design requirements. For example, the maximum rotor radius may have to be restricted to less than a pre-determined boundary to accommodate operational restrictions, hangar sizes, and air transportation (see Table 8.1). There is a fine art in balancing hover and forward flight power requirements.

The aerodynamic loads will then become input to the blade structural design that will aim to deliver an internal blade design capable to support the average, and cyclic loads on the structure as well as the centrifugal loads on the blades.

The structural design will lead to several interactions and iterations with the aerodynamic design and a modern trend is to exploit the bending and twisting of the blade to obtain better performance of the blade, while the blade is maintaining a safe distance from flutter and other limits, for example divergence. Estimation of the fatigue life of the blade should begin at this stage.

8.7 Fuselage Sizing

The fuselage will be sized according to payload requirements, as well as pilots cabin, passengers seats, flight systems and fuel tanks. The tail boom is part of the fuselage, alongside ancillary systems such as sponsons (for landing gear), engine bays, passengers and cargo doors.

Aside from the volumetric requirements, there will optimisation opportunities for the forebody and the aftbody, in order to reduce both hover and cruise drag. In hover, a wide fuselage creates a blockage to the rotor downwash; it is thus responsible for the so-called *vertical drag*, also discussed in §4.8. Therefore, it may be useful to set a limiter to the fuselage width, subject to load constraints.

The forebody will be optimised using advanced aerodynamic methods in order to reduce pressure drag. Likewise, the aft-body will be subject to large flow separations (inevitable), but it offers more opportunities for contouring, subject to weight limitations. Rotorcraft aftbodies come in great varieties: they can be fully streamlined, widely rounded, flat or even with blunted near-vertical shapes, Fig. 8.11.

8.8 Landing Gear Design

Helicopters have the ability to land almost anywhere, and this is facilitated by the variety of landing gear available. The set of options available includes landing skids, landing wheels (retractable or fixed), landing floats and even skis for operations over ice and snow. When the landing gear is made of fixed landing skids, there are

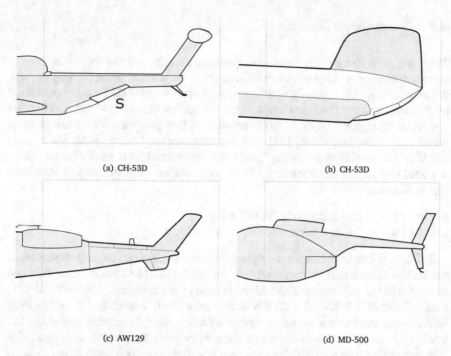

(a) CH-53D   (b) CH-53D

(c) AW129   (d) MD-500

Fig. 8.11 Aft fuselage geometries (not on the same scale). In graph (**a**), S denotes a splitter plate

generally two skids (with a nose pointing upward), two interconnecting tubes and some boarding steps.

In many cases, the under carriage units are not retractable and contribute to the total drag at all flight conditions. There are exceptions, such as the retractable landing gear of the AS332 Super Puma.

The main drag contributor is the exposed wheels, but struts and other geometrical details are also important. A typical under-carriage unit is made of one/two wheels on a single axle, one or two struts, and various cavities. For a two-wheel bogie, the wheel drag depends on the relative distance between wheels. The method of ESDU 79015 [19] is used.

The drag of the skids is calculated by using semi-empirical data for the inclined cylinder, following the method of ESDU [20]. The drag of the interconnecting tubes is split between: (a) sections below the airframe; (b) sections connecting the skids and the airframe. For the first component, the ESDU document provides data of lift and drag of cylinders in close proximity to a plane wall. The latter component is again an inclined cylinder.

8.9 Empennage Design

There is a great variety of helicopter empennage designs, and they have one thing in common: they are not symmetric with respect to the vertical plane passing through the longitudinal axis. No design guidelines are known to exist; this is partly explained by the peculiar aspects of the airflow in the aft portion of the rotorcraft, and the fact that there is a great variety of flight conditions that prevent generalisations from one design to another. Prouty [21] writes that neither the horizontal nor the vertical stabiliser "is absolutely necessary", although this statement is hardly useful. Thus, we need to consider again some top level requirements. For the vertical stabiliser, the requirements include:

- Provide structural support for the tail rotor
- Contribute to the directional stability of the vehicle.

It is noted that there is not even agreement on the definition of vertical stabiliser; the technical literature variously reports *fin*, *vertical tail* and other terms. In some cases, the tail rotor is mounted directly at the end of the tail boom (Leonardo AW139, Bell 206 and 407, MD-500, PZL SW-4, Robinson R44, Schweizer 330 and others). In this case, the vertical stabiliser has an added *ventral fin* running below the tail boom to protect the rotor at excessive pitch up rates on the ground. A *tail guard* is sometimes also present, as well as a separate vertical stabiliser, such as in the cases shown in Fig. 8.12.

As for the horizontal stabiliser, we have cases of single or twin stabilisers, e.g. on both sides of the tail boom. When a single horizontal surface is present, it is generally mounted on the lee side of the tail boom. This means that the retreating main rotor blade goes over the tail boom first (CH-53E Sea Stallion, AS332 Super

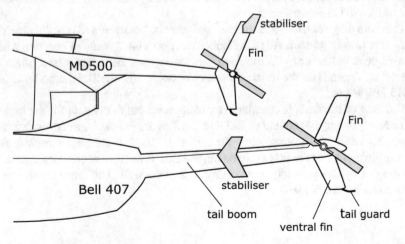

Fig. 8.12 Vertical stabiliser arrangements for the MD-500 (top) and the Bell 407 (bottom); the two rotorcraft are to the same scale

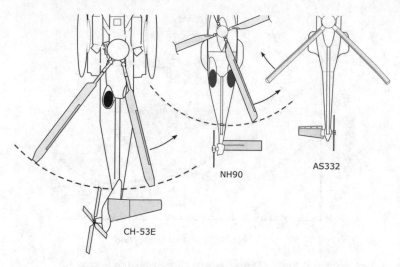

Fig. 8.13 Selected stabilisers with a single horizontal surface

Puma, the the NH90 and the Westland Lynx); this solution also prevents any risk of tail rotor blades colliding with the stabiliser. An example is shown in Fig. 8.13, where these rotorcraft are shown in the same scale. Note that in the case of the CH-53E the stabiliser is made of two segments, the inboard of which is swept upward $\sim 25°C$, and the outboard is essentially horizontal.

Where lift is generated by these stabilisers, it has a line of action that is offset on the longitudinal vertical plane of the rotorcraft. Hence this lift generates a rolling moment. Consider the CH-53E: the sense of rotation is anti-clockwise as seen from above. In advancing flight, the lift asymmetry would generate a rolling moment that would cause the rotorcraft to roll down on its left side. This is prevented by the rotor articulation of the cyclic pitch control through the swashplate.

The configuration aerodynamic methods presented in Chap. 4 for estimating steady-state aerodynamic response and corresponding derivatives, *if the inflow is known.* Unfortunately, for asymmetric cases where there is considerable swirl from the main rotor, the possibility of blade-vortex interaction, and flow separation instigated by the tail boom, linear aerodynamics is a poor approximation.

8.10   Aerodynamic Drag of the Helicopter

The elaboration of the aerodynamic drag is central to the quantification of power requirements, and hence the engine sizing. The absolute value of drag and the relative contribution depend on the specific rotorcraft and flight conditions. For a conventional helicopter, sufficiently *streamlined*, we would expect that in level flight the airframe contributes up to 40% of the total drag, the rotor head 20–40%, and the

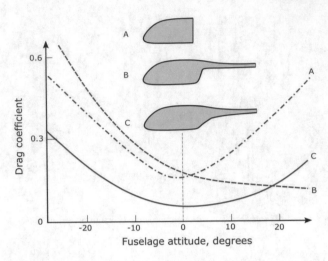

Fig. 8.14 Drag coefficient trends of fuselages with different aftbody shapes

remaining resistance coming from the tail rotor and the landing gear/skids. Abundant experimental data are available for clean configurations to provide terms of reference for high-fidelity aerodynamic simulations.

A low-order estimate is made with the method of components, wherein each separate contribution is added up, with limited or no consideration of interference effects. The problem is normally handled by considering an *equivalent flat plate area*, or scaling the total drag by the dynamic pressure, D/q.

Fuselage Drag. Since the fuselage is subject to a wide variety of angles of attack, the design must address drag, pitching moment and flutter effects over a variety of conditions; data exist in the technical literature that provide engineering information. In particular, the drag may or may not reach a minimum at zero-attitude, there can be a drop/increase in drag at some critical angles. Various devices have been used, such as vortex generators, flow splitters and flaps, each claiming some advantages. However, these solutions are very specific to the fuselage, and the results cannot always be extrapolated. Some examples are shown in Fig. 8.14. These data only apply to *clean fuselages* in isolation.

Fuselage drag will have to be augmented with the rotor hub drag, the landing gear (briefly discussed next) exposed surfaces and other ancillary systems specific to each rotorcraft.

Landing Skids Drag. The drag of the skids is calculated by using semi-empirical data for the inclined cylinder, following the method of ESDU [20]. The drag of the interconnecting tubes is split between: (a) sections below the airframe; (b) sections connecting the skids and the airframe. For the first component, the ESDU document provides data of lift and drag of cylinders in close proximity to a plane wall. The latter component is again an inclined cylinder.

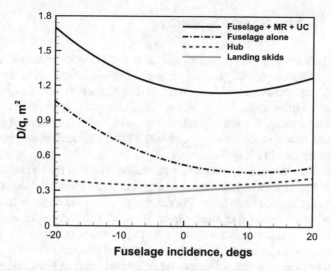

Fig. 8.15 Typical aerodynamic drag of a conventional helicopter with main contributions; MR = main rotor; UC = under-carriage/landing skids

The struts are considered as inclined circular cylinders with relatively high surface roughness. The particular case of an under-carriage made of a single wheel with a strut inclined by an angle φ on the vertical plane, the drag coefficient is estimated from

$$\frac{D}{q} = \left(\frac{C_D}{C_{D_o}}\right) C_{D_o}\,(wd_w) \tag{8.14}$$

where w is the width and d_w is the diameter of the wheel/tyre. The ratio C_D/C_{D_o} depends on both the aspect-ratio d_w/w and the Reynolds number. For ordinary values of d_w/w and sub critical Reynolds numbers, $Re < Re_c$, this quantity is virtually constant and equal to about 0.4. For super-critical Reynolds numbers, the effect of d_w/w is more pronounced.

A calculation of landing skids drag has been carried out for the Eurocopter EC-135 helicopter. This is $D/q \sim 0.55\,\mathrm{m}^2$ at $10\,\mathrm{m/s}$ and $D/q \sim 0.46\,\mathrm{m}^2$ at $30\,\mathrm{m/s}$, with the airframe at zero pitch angle.

Full Helicopter Drag. An example of aerodynamic drag for a conventional helicopter is shown in Fig. 8.15. The isolated fuselage has a behaviour like configuration C in Fig. 8.14. The analysis of the separate contribution provides a guideline of where to address drag issues.

8.11   Helicopter Engine Sizing

An essential element in the process of engine sizing is the definition of the overall
characteristics, which typically require the use of a single or a twin-engine rotorcraft.
Since rotorcraft manufacturers do not produce engines, these have to be contracted out
through a competitive adjudication, whose details cannot be described at this point.
It is sufficient to notice that the critical choice will be on the number of engines, on
the rated shaft power (uninstalled), on service availability and other characteristics
previously described in Chap. 3.

There is a need to exchange critical information between the engine and the airframe manufacturer, although the contract often requires confidentiality on both sides,
each having asymmetric knowledge. For example, the engine manufacturer does not
have detailed understanding of engine installation effects and engine-airframe integration issues. It is possible that a turboshaft already exists in the market that fulfills
key requirements. In this case, a derivative engine can be contracted out to deliver
specific power requirements, subject to other limitations (not least weight, particle
separators, support logistics) that will have to be assessed. A new turboshaft, a vari-
ant or otherwise, will have to be certified, which impacts on development costs. In
some cases, derivative engines can be found across the civil and military markets.

To start with, an assessment is carried out on the power requirements of the
rotorcraft using standard performance methods. A target shaft power is listed as
a derived user requirement, hence it is missing from the listing on Table 8.1. For
our purpose, an engine derivative among a family of existing engines may fit the
requirements.

8.11.1   Engine Power Requirements

We first estimate the rotorcraft power requirements in the most demanding conditions,
including MTOW, high altitude, high temperature, with the helicopter in hover. For
this purpose, we use the methods already described. In brief, the total power would be

$$P_{req} > P_{mr} + P_{tr} + P_a + P_{gbox} + P_{aux} + P_{climb} \tag{8.15}$$

where the main rotor power is described by Eq. 8.2, the tail rotor power is given
by Eq. 8.12, the airframe power in this instance is zero because the helicopter is in
hover; the transmission power P_{gbox} is given by an equation such as Eq. 3.1; finally,
P_{aux} is a power requirement to run auxiliary systems, although it can be delivered
by an APU.

Transmission limits will have to be established from the maximum design torque,
which will be a fixed percent higher than the limit engine torque. This is required to
guarantee safety and possibly also an engine upgrade at a later point, without having
to redesign the gearbox. Transmission architecture has been described in Chap. 3.

We need to select an excess specific power margin. This is set to 100 ft/minute *at the operational ceiling* z_p given in Table 8.1. Therefore:

$$\mathrm{SEP}(z_p) = \frac{P_{engine}(z_p) - P_{req}(z_p)/2}{W} = 100\,\text{feet/m} \qquad (8.16)$$

and the engine power at the operational ceiling becomes

$$P_{engine} = \frac{P_{req}}{2} + \mathrm{SEP} \cdot W \qquad (8.17)$$

Assuming in the first instance that the altitude power lapse follows the scaling of $\sim (p/p_o)^{0.8}$, we can infer the sea level power. Equation 8.17 could be a demand that cannot be met, because of costs or other engineering reasons; therefore, we may need to compromise and make less strict assumptions. These may include operating at altitudes on a normal day (instead of a hot day), or operating on a hot day at low altitudes, or impose take-off weight limitations at certain operational conditions. A sensitivity study at this point would allow fine tuning the power requirements in order to bring down weights and costs. For example, one option would be to have a turboshaft engine capable of delivering an emergency output power considerably larger than the continuous power. This feature would allow the engine size reduction without compromising flight safety [22].

The specific power P_{engine}/W can be compared with historical data to check that we are in line with prior art, but we do not necessarily need to stop there. In fact, another important consideration is the twin-engine operation. Recalling the OEI instances discussed in Chap. 4, the engine will have to be sized to deliver some excess power for a limited amount of time.

Example of Engine Sizing. The AgustaWestland (now Leonardo Helicopters Company) AW139 twin- engine helicopter is powered by Pratt& Whitney PT6C-67C turboshaft engines, whose ratings are available from their *Type Certificate*[1]. The maximum continuous power is 969 kW, and the OEI 2 and 1/2 minute power is 1217 kW (uninstalled). This implies a temporary power ramp up of over 25%. However, there is provision for a *continuous OEI* output of 1,064 kW, which is still 10% higher than the MCP. Since the total installed power would be 969 kW × 2, corresponding to a power loading $P/W = 0.277$ kW/kg, the emergency operation would have a good $P/W = 0.174$ kg/kW in the worst case scenario. This helicopter is relatively well powered as a twin-engine, since it is capable of high climb rates and fast level flight.

Statistical data for twin-engine helicopter power loading is shown in Fig. 8.16a. Note that the mean value is indicated by the horizontal dashed line: $P/W = 0.293$ kW/kg with a standard deviation of 0.032. Power ratings and corresponding nominal weights of in-service turboshaft engines are shown in Fig. 8.16b. Since these engines

---

[1] EASA Type Certificate Data Sheet IM.E.022, Issue 02, updated in Nov. 2012 (models PT6C-67C and -67E).

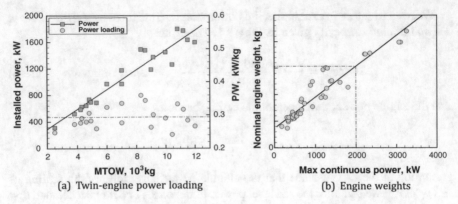

Fig. 8.16 Estimated power loadings for twin-engine helicopters (both civil and military)

are finely tuned to a specific rotorcraft, they may have different power ratings for nearly the same *nominal weight*. With regards to the latter term, the manufacturer must specify whether this is *dry weight* (e.g. without any residual fluids) and whether they include any ancillary systems. A turboshaft rated to an MCP of 2,000 kW is likely to have a dry weight of ~ 310 kg, corresponding to $P/W = 0.155$ kW/kg. For a twin-engine, this means $P/W = 0.31$ kW/kg, which is within one standard deviation of the mean value of the data displayed in Fig. 8.16a. When the engine weight is estimated, it can be used to estimate the rotorcraft structural weight.

References

1. Johnson W (2010) NDARC-NASA design and analysis of rotorcraft. Validation and demonstration. In: AHS Aeromechanics specialist conference
2. Weiand P, Krenik A (2018) A multi-disciplinary toolbox for rotorcraft design. Aeronautical J 122(1250):620–645. https://doi.org/10.1017/aer.2018.2
3. Martins J, Lambe A (2013) Multidisciplinary design optimization: a survey of architectures. AIAA J 51(9):2049–2075. https://doi.org/10.2514/1.J051895 Sept
4. Rand O, Khromov V (2002) Helicopter sizing by statistics. In: 58th AHS forum proceedings, Montreal, Canada
5. Sinsay JD (2018) Re-imagining rotorcraft advanced design. Aeronautical J 122(1256):1497–1521. https://doi.org/10.1017/aer.2018.107
6. Nickerson RS (1998) Confirmation bias: a ubiquitous phenomenon in many guises. Revier General Psychol 2(2):175–220
7. Moore MD, Fredericks B (2014) Misconception of electric propulsion aircraft and their emergent aviation markets. In: AIAA Scitech forum, AIAA 2014-0535. https://doi.org/10.2514/6.2014-0535
8. Filippone A, Barakos GN (2021) Rotorcraft systems for urban air mobility: a reality check. Aeronautical J 125(1283):3–21

9. Price M, Raghunathan S, Curran R (2006) An integrated systems engineering approach to aircraft design. Progr Aerosp Sci 42(4):331–376. https://doi.org/10.1016/j.paerosci.2006.11.002

10. Stepniewsky WZ, Keys CN (1984) Rotary wing aerodynamics. Dover Editions

11. Graham D, Sung D, Young L, Louie A, Stroub R (1989) Helicopter hub fairing and pylon interference drag. Technical Report TM-101052, NASA

12. Wiesner W, Kohler G (1974) Tail rotor design guide. Technical Report USAAMRDL Technical Report 73-99, Boeing Vertol Company

13. Kreitmar-Steck W, Hebensperger M (2016) Empennage of a helicopter. US Patent 9.266,602 B2, Airbus Helicopters DE GMBH

14. Mouille R (1970) The Fenestron, shrouded tail rotor of the SA. 341 Gazelle. J Am Helicopter Soc 15(4):31–37. https://doi.org/10.4050/JAHS.15.31

15. VanHorn J (1990) Circulation control slots in helicopter yaw control system. US Patent 4,948,068 A

16. Lynn RR, Robinson FD, Batra NN, Duhon JM (1970) Tail rotor design. Part I: aerodynamics. J Am Helicopter Soc 15(4):2–30. https://doi.org/10.4050/JAHS.15.4.2

17. Brocklehurst A, Barakos G (2013) A review of helicopter rotor blade tip shapes. Progr Aerosp Sci 56:35–74. https://doi.org/10.1016/j.paerosci.2012.06.003

18. Bousman WG (2003) Aerodynamic characteristics of SC1095 and SC1094-R8 airfoils. Technical Report TP-2003-212265, NASA

19. ESDU (1987) Undercarriage drag prediction methods. Data Item 79015. ESDU International, London

20. ESDU (1986) Mean forces, pressures and flow field velocities for circular cylinder structures: single cylinder with two-dimensional flow. Data item 80025. ESDU International, London

21. Prouty RW (1998) Helicopter performance, stability and control. Krieger Publ (reprint)

22. Hirschkron R, Haynes J, Goldstein D, Davis R (1981) Rotorcraft contingency power study. Technical Report NASA CR-174675, NASA

Antonio Filippone is at the School of Engineering, University of Manchester, where he specializes in aircraft and rotorcraft performance, environmental emissions aircraft noise, and engine performance.

George Barakos is at the School of Engineering, Glasgow University, where he specialises in high-fidelity aerodynamics and aero-acoustics models, computational fluid dynamics, high-performance computing and rotorcraft engineering.

Appendix A
Numerical Solution of Acoustic Problems

In Chap. 6, in particular in § 6.4.3, we described how the Farassat 1A formulation can be implemented as an algorithm to compute the rotor blade loading and thickness noise. In this section, we take this further and present MATLAB code examples to compute the loading and thickness noise of a rotor blade over a single rotation. We will take as input the blade kinematics and loading over a rotation and output the acoustic pressure over that rotation. Whilst the code does not necessarily represent the most comprehensive acoustic analysis, it should provide the reader with an understanding on how to construct a code for the acoustic analysis of rotorcraft and also serve as a useful starting point for the reader to extend and apply to their own particular applications.

We will use a mid-panel quadrature and source-dominant implementation. For a list of symbols, please refer back to Chap. 6. We will assume that the blade has been discretised into a number of chordwise and spanwise panels and that the corresponding kinematics and aerodynamics have been computed by another programme. The code will be broken into a number of smaller sections in order to assist the reader in understanding a number of important steps in the development of the code. However, when the sections are brought together, they will form a working code for the acoustic analysis of arbitrary rotors.

Before we begin building up the rotor noise code, we will first take a step back and look at the steps required to build the code. A top-level description of the code is shown in Fig. A.1. We will use this as a guide and refer back to it as we build up each element of the code. Each block in the diagram generally represents a single section of the code. Therefore, with the two components, the reader should be able to construct the full working code from the example codes and the figure below.

With reference to Fig. A.1, we can see the first step is to load in the input files. In the present example, we will have two input files. The first input file will describe the kinematics and loading for the discretised blade over a single rotation. In the present example we assume that the data is nested firstly in the chordwise direction, secondly in the spanwise direction, and finally in time. The format of the file is as follows:

A. Filippone and G. Barakos (eds.), *Lecture Notes in Rotorcraft Engineering*, Springer Aerospace Technology, https://doi.org/10.1007/978-3-031-12437-2

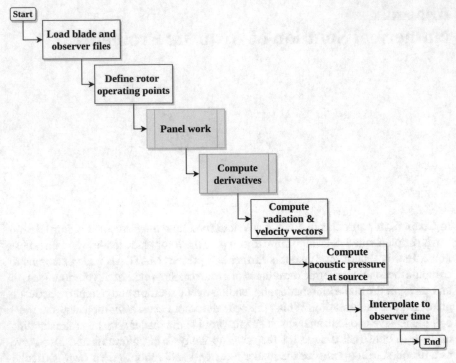

Fig. A.1 Top-level description of Farassat-1A code implementation

$$
\begin{array}{cccc}
N_{time} & N_{span} & N_{chord} & 0 \\
x_{1,1,1} & y_{1,1,1} & z_{1,1,1} & p_{1,1,1} \\
x_{2,1,1} & y_{2,1,1} & z_{2,1,1} & p_{2,1,1} \\
\vdots & \vdots & \vdots & \vdots \\
x_{N_{chord},1,1} & y_{N_{chord},1,1} & z_{N_{chord},1,1} & p_{N_{chord},1,1} \\
x_{1,2,1} & y_{1,2,1} & z_{1,2,1} & p_{1,2,1} \\
x_{2,2,1} & y_{2,2,1} & z_{2,2,1} & p_{2,2,1} \\
\vdots & \vdots & \vdots & \vdots \\
x_{N_{chord},N_{span},1} & y_{N_{chord},N_{span},1} & z_{N_{chord},N_{span},1} & p_{N_{chord},N_{span},1} \\
x_{1,1,2} & y_{1,1,2} & z_{1,1,2} & p_{1,1,2} \\
x_{2,1,2} & y_{2,1,2} & z_{2,1,2} & p_{2,1,2} \\
\vdots & \vdots & \vdots & \vdots \\
x_{N_{chord},1,2} & y_{N_{chord},1,2} & z_{N_{chord},1,2} & p_{N_{chord},1,2} \\
\vdots & \vdots & \vdots & \vdots \\
x_{N_{chord},N_{span},2} & y_{N_{chord},N_{span},2} & z_{N_{chord},N_{span},2} & p_{N_{chord},N_{span},2} \\
\vdots & \vdots & \vdots & \vdots \\
x_{N_{chord},N_{span},N_{time}} & y_{N_{chord},N_{span},N_{time}} & z_{N_{chord},N_{span},N_{time}} & p_{N_{chord},N_{span},N_{time}}
\end{array}
$$

The first row in the input file contains the details of the input file. Firstly, the number of time steps, spanwise points and chordwise points are given. The first row is trailed by a 0 to preserve the row and column size. As we will see later, these values are important. The number of time points will tell us the azimuthal step size of the simulation. Along with the blade discretisation, these values will assist in extracting the data from the file. The file proceeds in the nested order of chordwise, spanwise and time points of the discretised blade coordinates and the surface pressure at each point.

The second input file contains the observer coordinates. This is a simpler file with only three columns describing the coordinates of the observer:

$$
\begin{array}{ccc}
x_1 & y_1 & z_1 \\
x_2 & y_2 & z_2 \\
x_{N_{obs}} & y_{N_{obs}} & z_{N_{obs}}
\end{array}
$$

where N_{obs} is the number of observer points and x, y, z are the Cartesian coordinates of the observer. We must pause to note that care must be taken such that the observer and source coordinates are correctly specified. For example, if the rotor blade is at an altitude of 304.8 m (1000 feet) it may be necessary to add this to the source coordinates as the source is likely modelled relatively, with atmospheric properties used to account for the effects of altitude. Alternatively, we may subtract the value from the observer coordinates. The important point is that the relative distance and direction between source and observer is captured.

Having defined the assumed format for the input files, we can load these into our MATLAB script, Listing A.1.

The first variable, $\texttt{dirin}$, specifies the location of the input files for the particular run. We then first focus on extracting the blade kinematics and loading data. Using the first line of the input file, we gather details of the file format, lines 8–10. Following pre-allocation of the source coordinates and pressure variables, lines 14–15, we use these variables to extract the data from the file in the nesting order we have described (chord-span-time). Knowing the number of time points and spanwise points allows us to set up the loops over the data and finally the number of chordwise points allows us to extract the data without a further loop over the data. The remainder of listing extracts the observer data from it's corresponding input file. Note here that we have assumed all points are given in metres and Pascals.

Referring back to our code outline, Fig. A.1, the next step is definition of the rotor operating point. Whilst this could be contained within an input file (for which the user can reference our previous step for guidance), in the present case, we simply define the operating points within the script. The code with the required variable declarations is shown in Listing A.2, where the comments in the Listing shows the assumed units for each variable. In this example, we have defined the rotor velocity (e.g. climbing flight), rotational speed, atmospheric properties, and the blade count. Note in the Listing, $\texttt{<val>}$ should be replaced by the appropriate value.

We next move on to the section described as 'Panel Work'. In our input data we have assumed each point corresponds to a panel vertex and since we have opted to

Listings A.1: Blade kinematics and loading input file

```
1   % working directory:
2    dirin = 'location_of_data';
3
4   % Kinematics and loading data:
5    data1 = importdata([dirin,'/loading_file']);
6
7   % Get details from first row of data:
8    Nt      = data1(1,1);          % Number of time points
9    Nslices = data1(1,2);          % Number of spanwise points
10   Nchord  = data1(1,3);          % Number of chordwise points
11
12   % Extract panel coordinates and pressure at each time step:
13   % pre−allocate:
14   X = zeros(Nslices,Nchord,Nt); Y = zeros(Nslices,Nchord,Nt); Z =
15   zeros(Nslices,Nchord,Nt); P = zeros(Nslices,Nchord,Nt);
16
17   % loop over time
18   for i = 1:Nt
19      % loop over spanwise points
20      for ii = 1:Nslices
21          % can work out chordwise points
22          % plus two as first line is format data
23          lower = (i−1)*(Nslices*Nchord)+(ii−1)*Nchord+2;  % lower chord data
24          upper = lower+Nchord−1;                          % upper chord data
25          X(ii,:,i) = data1(lower:upper,1);       % X source coord
26          Y(ii,:,i) = data1(lower:upper,2);       % Y source coord
27          Z(ii,:,i) = data1(lower:upper,3);       % Z source coord
28          P(ii,:,i) = data1(lower:upper,4);       % Blade surf. pres.
29      end
30   end
31
32   % observer location file:
33   data2 = importdata([dirin,'/obs_file']);
34
35   % extract observer locations
36   xobs = data2(:,1); yobs = data2(:,2); zobs = data2(:,3);
```

Listings A.2: Rotor operating point defintion

```
1   Vx  = <val>;     %[m/s] x−comp. free−stream velocity
2   Vy  = <val>;     %[m/s] y−comp. free−stream velocity
3   Vz  = <val>;     %[m/s] z−comp. free−stream velocity
4   RPM = <val>;     %[rev/min] rotor rotational speed
5   c0  = <val>;     %[m/s] sound speed at alt.
6   rho = <val>;     %[kg/m^3] density at alt.
7   N   = <val>;     %[−] Rotor blade count
```

Fig. A.2 Arbitrary panel

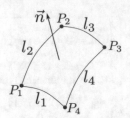

use a mid-panel quadrature algorithm, we must approximate the panel centroids as well as the pressure at this point. Further, the use of this approach requires the panel area which can be easily computed from the panel data. Finally, inspection of the equations to compute the acoustic pressure, Eqs. (6.32) and (6.33) shows we require velocity components in the normal directions. Therefore, it makes sense to compute the panel normals whilst we are working with the panels themselves. All these tasks are wrapped in a function called `Panel`, Listing A.3.

The `Panel` function takes as input the blade coordinates and the pressure at each point, and returns the panel centroids, the pressure at the centroids, the unit normal at each centroid and the panel areas. Figure A.2 shows an arbitrary panel that may be found on the blade, defined by the four points P_1–P_4. The panel centroids are calculated from the points defining each panel:

$$P_c = \frac{1}{4} (P_1 + P_2 + P_3 + P_4) \tag{A.1}$$

This computation is shown in Lines 20–22 of Listing A.3 for the three cooridinate components. The reader should note the variables `IP1` and `IP2` which are introduced in Lines 9–17 to ensure the correct orientation for the panels for outward normals. The reader is recommended to verify this using a `quiver` plot of the normals.

The unit normals are computed accoriding to:

$$\vec{n} = \frac{L_u \times L_v}{\|L_u \times L_v\|} \tag{A.2}$$

as shown in Line 41. From Fig. A.2, $L_u = \frac{1}{2} (l_1 + l_3)$ and $L_v = \frac{1}{2} (l_2 + l_4)$ are the average opposing side lengths of each panel, Lines 27–33. These variables are also then used in the computation of the panel area, Line 45, according to:

$$A_p = \|L_u \times L_v\| \tag{A.3}$$

Listing A.4 shows the call to the `Panel` function for each time step and the important pre-allocation of variables prior to entering the loop.

Listings A.3: `Panel` function

```
1   function[Xc,n,Pc,Ap] = Panel(X,Y,Z,P)
2   % Pre-allocate:
3   N1 = size(X,1); N2 = size(X,2); isign = zeros(N1-1,N2-1); Xc =
4   zeros(N1-1,N2-1,3); Pc = zeros(N1-1,N2-1); Lu = Xc; Lv = Xc;
5
6   for i = 1:N1-1
7       for ii = 1:N2-1
8           % Check for upper lower surface:
9           if ii<N2/2  % lower
10              IP1 = ii; IP2 = ii+1;   isign(i,ii) = -1;
11          end
12          if ii>N2/2+1 % Upper:
13              IP1 = ii+1; IP2 = ii;
14          end
15          if ii>N2/2
16              isign(i,ii) = +1;
17          end
18
19          % Panel centroid:
20          Xc(i,ii,1) = 0.25*(X(i,IP1)+X(i+1,IP1)+X(i+1,IP2)+X(i,IP2));
21          Xc(i,ii,2) = 0.25*(Y(i,IP1)+Y(i+1,IP1)+Y(i+1,IP2)+Y(i,IP2));
22          Xc(i,ii,3) = 0.25*(Z(i,IP1)+Z(i+1,IP1)+Z(i+1,IP2)+Z(i,IP2));
23
24          % Pressure at centroid
25          Pc(i,ii) = 0.25*(P(i,IP1)+P(i+1,IP1)+P(i+1,IP2)+P(i,IP2));
26
27          Lu(i,ii,1) = 0.25*(X(i,IP1)+X(i+1,IP1)-X(i+1,IP2)-X(i,IP2));
28          Lu(i,ii,2) = 0.25*(Y(i,IP1)+Y(i+1,IP1)-Y(i+1,IP2)-Y(i,IP2));
29          Lu(i,ii,3) = 0.25*(Z(i,IP1)+Z(i+1,IP1)-Z(i+1,IP2)-Z(i,IP2));
30
31          Lv(i,ii,1) = 0.25*(-X(i,IP1)+X(i+1,IP1)+X(i+1,IP2)-X(i,IP2));
32          Lv(i,ii,2) = 0.25*(-Y(i,IP1)+Y(i+1,IP1)+Y(i+1,IP2)-Y(i,IP2));
33          Lv(i,ii,3) = 0.25*(-Z(i,IP1)+Z(i+1,IP1)+Z(i+1,IP2)-Z(i,IP2));
34      end
35  end
36
37  LuCOV = P1./((Lu(:,:,1).^2+Lu(:,:,2).^2+Lu(:,:,3).^2).^0.5); LvCOV =
38  isign.*Lv./((Lv(:,:,1).^2+Lv(:,:,2).^2+Lv(:,:,3).^2).^0.5);
39
40  % Unit normal
41  n = cross(LuCOV,LvCOV,3);
42
43  % Panel area:
44  c1 = cross(Lu,Lv,3); Ap =
45  (c1(:,:,1).^2+c1(:,:,2).^2+c1(:,:,3).^2).^0.5;
46  end
```

The next step in the process is the computation of variable derivatives. The corresponding code is presented in Listing A.5. The first step is to define the array of

Listings A.4: Blade panel work

```
 1   % Get panel centroid data from input data using PANEL function
 2   Xc = zeros(Nslices−1,Nchord−1,3,Nt);      % [m] Panel centroid coordinates
 3   n = Xc;                                     % [−] Unit normal at each centroid
 4   Pc = zeros(Nslices−1,Nchord−1,Nt);         % [Pa] Pressure at centroid
 5   Ap = zeros(Nslices−1,Nchord−1);            % [m^2] Panel area
 6
 7   % Call the function
 8   for i = 1:Nt
 9       [Xc(:,:,:,i),n(:,:,:,i),Pc(:,:,i),Ap] = Panel(squeeze(X(:,:,i)),...
10       squeeze(Y(:,:,i)),squeeze(Z(:,:,i)),squeeze(P(:,:,i)));
11   end
```

source points. The derivatives will be computed with respect to the source time, therefore, we first define the azimuthal discretisation, where we have assumed the input data is for a single rotation from 0 to 360°, from which we can compute the source emission time for each panel from the rotational speed. This is the stage of the code where we incorporate the elements of the source-dominant algorithm. In particular, we are setting the source time array for each panel. If we were to utilise an observer-dominant algorithm, this would be the stage where we would declare the observer time and compute the corresponding source emission times.

Listings A.5: Computing variable derivatives

```
 1   % Get azimuthal discretisation
 2   % we assume that data is for a single revolution and is equally spaced
 3   psi = linspace(0,2*pi,Nt);        % [rad] Azimuth angle
 4   te  = psi./(abs(RPM)*2*pi/60);    % [s] Time discretisation
 5
 6   % Get velocity of each panel using central differences:
 7   Vp = Cdiff(Xc,te);
 8   % add free−stream components
 9   Vp(:,:,1,:) = Vp(:,:,1,:)+Vx; Vp(:,:,2,:) = Vp(:,:,2,:)+Vy;
10   Vp(:,:,3,:) = Vp(:,:,3,:)+Vz;
11
12   % Get acceleration of each panel using central differences
13   Vd = Cdiff(Vp,te);
14
15   % Take first derivative of unit normal:
16   nd = Cdiff(n,te);
17
18   % Get pressure derivative using central differences:
19   pdot = Cdiff(Pc,te);
```

The next step is to perform the actual derivative computation. From Eqs. (6.32) and (6.33), we can see we need to compute the derivatives for the variables M_n, M_r and p_S. From Listing A.5, we first compute the derivative of the positon to get the blade velocity and add to this the free-stream variables. We then differentiate these velocities to get the blade accelerations. Using the sound speed these will be used to compute the Mach number components at a later stage. We then compute the derivative of the unit normal and the blade surface pressure. Due to the repeated calculations, the derivative computation is wrapped within a function, `CDiff`, which is shown in Listing A.6. The function takes as input the variable of interest and the time discretisation and outputs the derivative computed using central differences. We have a number of notable points in the `Cdiff` function. Firstly, we introduce the variable `otherdims`, Line 10, which allows us to handle variables of differeing dimensions, so long as the time component is on the last dimensions. The next notable point is that we have to have take special care for the first and last time points in the variable, lines 17 and 19.

Listings A.6: `Cdiff` function

```
1   function[deriv] = Cdiff(var,t)
2   % Function to compute central differences of input array
3   % Inputs : variable (size Nslices x Nchord) and index
4   % Outputs: derivative of variable at index
5
6   % pre-allocate vars
7   deriv = zeros(size(var));
8
9   % Get number of dimensions allows us to use different sizes
10  otherdims = repmat({':'},1,ndims(deriv)-1);
11
12  % get last index:
13  endx = length(var);
14
15  % Use forward differences for first and last points
16  % avoids issues if psi(1) == psi(end);
17  deriv(otherdims{:},2:endx-1) =
18  (var(otherdims{:},3:endx)-var(otherdims{:},1:endx-2))./(2.*(t(2)-t(1)));
19  deriv(otherdims{:},1) =
20  (var(otherdims{:},1)-var(otherdims{:},2))./(t(1)-t(2));
21  deriv(otherdims{:},endx) =
22  (var(otherdims{:},endx)-var(otherdims{:},endx-1))./(t(2)-t(1)); end
```

The next step in the development of the code is the computation of the radiation and velocity vectors. This part is where we compute the majority of the variables that populate the equations for the acoustic pressure. The corresponding code is shown in Listing A.7. The first important aspect to note in the code example is that this is the point where we enter the loops for the observer and the source time. From this point we are calculating variables, and subsequently the acoustic pressure, for each observer and each source time. The computed values will be appended to variables in order to get the time pressure histories for each observer. In order to improve the memory efficiency of this step, the required variables should be pre-allocated prior to values being appended. After extracting the current iteration of observer point, for each source time we first compute the radiation vector, shown on Lines 11–13, according to:

$$\vec{r} = x_{obs} - \vec{X}_c \tag{A.4}$$

where, $\vec{X}_c$ is the source coordinates, here the panel centroids. We then compute the magnitude for each panel at the current source time in order to calculate the unit radiation vector, Line 19. The user should note that, unlike all other variables in the present listing, we save the value of the magnitude of the radiation vector at each iteration of source time, Line 20. Therefore, it is good practice to pre-allocate this variable. The reason for saving this variable after each iteration will become clear as we progress in the code develoment.

We next compute the velocity and Mach number projections. To illustrate this step, we will demonstrate the computation of just a few of these variables. For example, the projection of the velocity in the normal direction is computed according to:

$$V_n = \vec{V}_p \cdot \vec{n} \tag{A.5}$$

where, $\vec{V}_p$ and $\vec{n}$ are the panel velocty and unit normal components. The derivative of this variable is then given as:

$$V_{nd} = \vec{V}_d \cdot \vec{n} + \vec{n}_d \cdot \vec{V}_p \tag{A.6}$$

where, V_d and n_d are respectively the first derivative of the velocity and unit normal with respect to the source time. The Mach number projections and their corresponding derivatives are computed in a similar manner as shown in Lines 30–39 of Listing A.7. Following the computation of the Mach number projection, we compute the cosine of the angle between the normal and radiation directions:

$$\cos \Theta = \vec{r} \cdot \vec{n} \tag{A.7}$$

The corresponding derivative is computed similarly to that of the velocity and Mach number variables. Finally, we compute the familiar Doppler factor: $1 - M_r$. Thus we now have all the variables required to compute the acoustic pressure for our rotor.

We are now approaching the end of our code example, such that we are ready to compute the acoustic pressure due to thickness and loading at the source. The code for this computation is shown in Listing A.8 and reflects Eqs. (6.32) and (6.33) for the thickness and loading components respectively. In the present example we have broken the equations down in order to improve readibility. However, the components should be readily identifiable from the corresponding equations. The reader should recall that we are using the mid-panel quadrature algorithm, Eq. (6.36) and is why we are multiplying by panel area when we bring all the thickness and loading components together, Lines 11 and 25 respectively. It should also be noted at the end of this code block we see the end for the source time loop. Therefore, the process is repeated for every observer time and we store the thickness and loading pressure of each panel at each source in the variables ppt and ppl.

At present we have the acoustic pressure for each panel at each source time. Therefore, in our final step we must compute the acoustic pressure that the observer actually hears. For this, we have to compute the time at which each source reaches the observer. Further, we would, instead of the rotor noise being described by an array of panel sources, like it to be represented as a single source in time. Therefore, we must interpolate each source onto a common observer time array—this forms the second part of the source dominant algorithm. Following this, we can perform the integration and represent the rotor blade as a single source in time. The code required to carry out this operation is shown in Listing A.9

As we have disucssed, the first step is to compute the time at which each source reaches the observer. We first do this by expanding our source time array to have the first two dimensions the same as our blade discretisation, Lines 7–8. Note for each panel the source time is the same. We then compute the time each source reaches the observer, according to:

$$t = \tau + R_i/c_0 \tag{A.8}$$

The reader should note that R_i is the magnitude of the radiation vector, the variable that we saved at each time step in Listing A.7. The next step is to create a new array of observer time. This will be the common array for which each source will be interpolated onto. This new array should cover as best as possible the extremes of the initial observer time array. To achieve this, we compute the average of the two extremes as our starting point, Line 15, and add to it the time for a single rotation, Line 17, and create an even distibution of points between these extremes. Lines 22–27 show the loops over each panel to interpolate from the source-observer time to the new common observer array. Note we are using linear interpolar and extrapolating any outlying points. With each source now computed at the same observer time array, we can perform the surface integration, here a summation, to represent the rotor blade as a single noise source, Lines 30 and 31 for the thickness and loading source respectively. The final line of the listing shows the end representing the end of the loop over the observer points.

To conclude, the rationale of the present example was to breakdown the process of developing a simple code for the reader to make a start in developing their own rotor codes. Combining the code examples provided in the text should make a complete and working rotor noise code. Whilst the presented code provides only a simple example, it can be easily built upon to improve its robustness, accuracy and its application to a wider range of cases.

Listings A.7: Computation of radiation and velocity vectors

```
1    % Loop over each observer
2    for i = 1:length(xobs)
3        % Get current observer positions
4        xo = [xobs(i),yobs(i),zobs(i)];
5
6        % Now loop over each azimuth/time step
7        for ii = 1:length(psi)
8
9            % Compute radiation vector
10           % Difference between source and observer:
11           r(:,:,1) = (xo(1)-squeeze(Xc(:,:,1,ii)));
12           r(:,:,2) = (xo(2)-squeeze(Xc(:,:,2,ii)));
13           r(:,:,3) = (xo(3)-squeeze(Xc(:,:,3,ii)));
14
15           % Compute the magnitude:
16           R = (r(:,:,1).^2+r(:,:,2).^2+r(:,:,3).^2).^0.5;
17
18           % Unit radiation vector:
19           rR = r./R;
20           Ri(:,:,ii) = R;
21
22           % Projection of velocity in normal direction:
23           Vn = squeeze(dot(squeeze(Vp(:,:,:,ii)),squeeze(n(:,:,:,ii)),3));
24
25           % Derivative of normal velocity:
26           Vnd = squeeze(dot(squeeze(Vd(:,:,:,ii)),squeeze(n(:,:,:,ii)),3)+dot(
                     squeeze(nd(:,:,:,ii)),squeeze(Vp(:,:,:,ii)),3));
27
28
29           % Mach number projected in radiation direction:
30           Mr = squeeze(dot(squeeze(Vp(:,:,:,ii)),rR,3))./c0;
31
32           % First derivative of Mach number in radiation direction:
33           Mrd = squeeze(dot(squeeze(Vd(:,:,:,ii)),rR,3))./c0;
34
35           % Absolute Mach number:
36           Ma = ((Vp(:,:,1,ii).^2+Vp(:,:,2,ii).^2+Vp(:,:,3,ii).^2).^0.5)./c0;
37
38           % Mach number in normal direction:
39           Mn = Vn./c0;
40
41           % Cos(theta):
42           cstheta = (dot(rR,squeeze(n(:,:,:,ii)),3));
43           % And it's derivative:
44           csthetad = (dot(rR,squeeze(nd(:,:,:,ii)),3));
45
46           % Doppler factor
47           dopp = 1-Mr;
```

Listings A.8: Computation of the acoustic pressure.

```
1    % Thickness source (Equation (6.32) ):
2
3    % Near−field :
4    ptn1 = rho∗c0.∗Vn.∗(Mr−Ma.^2)./(R.^2.∗dopp.^3);
5
6    % far−field
7    ptf1 = (rho.∗Vnd)./(R.∗dopp.^2);
8    ptf2 = (rho.∗Vn.∗Mrd)./(R.∗dopp.^3);
9
10   % Bring terms together and append to variable
11   ppt(:,:,ii) = (1/(4∗pi)).∗(ptn1+ptf1+ptf2).∗Ap;
12
13
14   % Loading noise (Equation(6.33)):
15
16   % Near−field :
17   pln1 = squeeze(Pc(:,:,ii)).∗(cstheta−Mn)./(R.^2.∗dopp.^2);
18   pln2 = squeeze(Pc(:,:,ii)).∗cstheta.∗(Mr−Ma.^2.)./(R.^2.∗dopp.^3);
19
20   %far−field :
21   plf1 = (pdot(:,:,ii).∗cstheta+Pc(:,:,ii).∗csthetad)./(c0.∗R.∗dopp.^2);
22   plf2 = Mrd.∗Pc(:,:,ii).∗cstheta./(c0.∗R.∗dopp.^3);
23
24   % Bring terms together and append to variable
25   ppl(:,:,ii) =  (1/(4∗pi)).∗(plf1+plf2+pln1+pln2).∗Ap;
26
27 end % End of azimuth loop
```

Listings A.9: *Interpolation onto observer time.*

```
1      %% Interpolate to observer time
2      % We currently have loading and thickness sources at source time
3      % We need to create an observer array and to interpolate onto this
4
5      % First compute time that source reaches observer:
6      % reshape our azimuth time to a time for each panel
7      tau = reshape(te,1,1,length(te));
8      tau = repmat(tau,Nslices-1,Nchord-1,1);
9      t = tau+Ri./c0;        % Time of each panel reaches observer
10
11     % Create array of observer time:
12     % This will be different for each observer
13     % Can make a global one, but caution needed not to lose sources
14     % Get the average arrival time at psi = 0
15     to1 = 0.5.*(min(min(Ri(:,:,1)))+max(max(Ri(:,:,1))))./c0;
16     % add time of one revolution to this:
17     to2 = to1+1/(abs(RPM)/60);
18     % create array:
19     tobs = linspace(to1,to2,Nt);
20
21     % Now interpolate each spanwise and chordwise panel:
22     for iii = 1:Nslices-1
23         for iv = 1:Nchord-1
24             ppl2(iii,iv,:) = interp1(squeeze(t(iii,iv,:)),squeeze(ppl(iii,iv,:))
                   ,tobs,'linear','extrap');
25             ppt2(iii,iv,:) = interp1(squeeze(t(iii,iv,:)),squeeze(ppt(iii,iv,:))
                   ,tobs,'linear','extrap');
26         end
27     end
28
29     % Now perform integration (sum)
30     pt = squeeze(sum(sum(ppt2,2),1));
31     pl = squeeze(sum(sum(ppl2,2),1));
32
33  end %end of observer loop
```

Index

Printed in the United States
by Baker & Taylor Publisher Services